PIPELINE SAFETY

HEARING

BEFORE THE

SUBCOMMITTEE ON ENERGY AND POWER

OF THE

COMMITTEE ON ENERGY AND COMMERCE HOUSE OF REPRESENTATIVES

ONE HUNDRED THIRD CONGRESS

SECOND SESSION

ON

H.R. 4394 and H.R. 4616

BILLS TO ESTABLISH ONE-CALL SYSTEMS TO IMPROVE NATURAL GAS AND HAZARDOUS LIQUID PIPELINE SAFETY, AND FOR OTHER PURPOSES

JUNE 23, 1994

Serial No. 103–134

Printed for the use of the Committee on Energy and Commerce

U.S. GOVERNMENT PRINTING OFFICE

84-088CC WASHINGTON : 1994

For sale by the U.S. Government Printing Office

Superintendent of Documents, Congressional Sales Office, Washington, DC 20402

ISBN 0-16-046105-7

COMMITTEE ON ENERGY AND COMMERCE

JOHN D. DINGELL, Michigan, *Chairman*

HENRY A. WAXMAN, California
PHILIP R. SHARP, Indiana
EDWARD J. MARKEY, Massachusetts
AL SWIFT, Washington
CARDISS COLLINS, Illinois
MIKE SYNAR, Oklahoma
W.J. "BILLY" TAUZIN, Louisiana
RON WYDEN, Oregon
RALPH M. HALL, Texas
BILL RICHARDSON, New Mexico
JIM SLATTERY, Kansas
JOHN BRYANT, Texas
RICK BOUCHER, Virginia
JIM COOPER, Tennessee
J. ROY ROWLAND, Georgia
THOMAS J. MANTON, New York
EDOLPHUS TOWNS, New York
GERRY E. STUDDS, Massachusetts
RICHARD H. LEHMAN, California
FRANK PALLONE, JR., New Jersey
CRAIG A. WASHINGTON, Texas
LYNN SCHENK, California
SHERROD BROWN, Ohio
MIKE KREIDLER, Washington
MARJORIE MARGOLIES-MEZVINSKY, Pennsylvania
BLANCHE M. LAMBERT, Arkansas

CARLOS J. MOORHEAD, California
THOMAS J. BLILEY, JR., Virginia
JACK FIELDS, Texas
MICHAEL G. OXLEY, Ohio
MICHAEL BILIRAKIS, Florida
DAN SCHAEFER, Colorado
JOE BARTON, Texas
ALEX MCMILLAN, North Carolina
J. DENNIS HASTERT, Illinois
FRED UPTON, Michigan
CLIFF STEARNS, Florida
BILL PAXON, New York
PAUL E. GILLMOR, Ohio
SCOTT KLUG, Wisconsin
GARY A. FRANKS, Connecticut
JAMES C. GREENWOOD, Pennsylvania
MICHAEL D. CRAPO, Idaho

ALAN J. ROTH, *Staff Director and Chief Counsel*
DENNIS B. FITZGIBBONS, *Deputy Staff Director*
MARGARET A. DURBIN, *Minority Chief Counsel and Staff Director*

SUBCOMMITTEE ON ENERGY AND POWER

PHILIP R. SHARP, Indiana, *Chairman*

EDWARD J. MARKEY, Massachusetts
RICHARD H. LEHMAN, California
CRAIG A. WASHINGTON, Texas
MIKE KREIDLER, Washington
BLANCHE M. LAMBERT, Arkansas
AL SWIFT, Washington
MIKE SYNAR, Oklahoma
W.J. "BILLY" TAUZIN, Louisiana
RALPH M. HALL, Texas
RICK BOUCHER, Virginia
JIM COOPER, Tennessee
JOHN D. DINGELL, Michigan (Ex Officio)

MICHAEL BILIRAKIS, Florida
J. DENNIS HASTERT, Illinois
CLIFF STEARNS, Florida
PAUL E. GILLMOR, Ohio
SCOTT KLUG, Wisconsin
GARY A. FRANKS, Connecticut
MICHAEL D. CRAPO, Idaho
CARLOS J. MOORHEAD, California (Ex Officio)

SHELLEY N. FIDLER, *Staff Director*
JUDI GREENWALD, *Special Assistant*
JUDITH A. O'BRIEN, *Special Assistant*
CATHERINE G. VAN WAY, *Minority Counsel*

CONTENTS

PIPELINE SAFETY

THURSDAY, JUNE 23, 1994

HOUSE OF REPRESENTATIVES,
COMMITTEE ON ENERGY AND COMMERCE,
SUBCOMMITTEE ON ENERGY AND POWER,
Washington, DC.

The subcommittee met, pursuant to notice, at 10 a.m., in room 2322, Rayburn House Office Building, Hon. Philip R. Sharp (chairman) presiding.

Mr. SHARP. The subcommittee will be in order, please.

Exactly 3 months ago the lives of the residents of the Durham Woods Apartment Complex in Edison, New Jersey changed forever. It was on March 23, 1994, that a 36-inch transmission pipeline located about 300 yards from this apartment complex ruptured. Since that time, there have been two other pipeline explosions, one in Green River, Wyoming and the other in Allentown, Pennsylvania. Both of these accidents resulted in the loss of life and property.

Many people are asking, "How did these accidents happen and what can we do in the future to prevent such tragedies from ever happening again?" Our goal today is to explore these questions and to discuss specific legislative initiatives that have been introduced in an attempt to provide us with answers.

The hearing will focus on two pieces of legislation that originated from a collaborative effort by some members. The bills are H.R. 4394, the Comprehensive One-call Notification Act of 1994 introduced by Congressman Pallone and by Senator Bradley—S. 2101—on May 11, and H.R. 4616, the Natural Gas Pipeline Safety Improvement Act of 1994, which was just introduced on Tuesday by Congressman Pallone and by Senators Bradley and Lautenberg.

H.R. 4394 addresses the need for a comprehensive nationwide one-call notification system, which is designed to help prevent the occurrence of damage to underground facilities such as pipelines and electric lines and to prevent injury or death to the excavator performing the work. We will also be discussing the amendment in the nature of a substitute that I intend to offer at the subcommittee markup. The substitute represents an additional effort by interested parties to reach consensus on the legislation in hopes of moving it before the end of this session. H.R. 4616 addresses some pipelines issues that have become of concern in light of the Edison explosion.

Our witnesses today include representatives from many organizations. Because the one-call legislation is comprehensive, meaning that all underground facility operators, as well as all excavators are required to participate in their State's one-call system, we will be hearing many witnesses today and many more are submitting testimony for the record. We will also be hearing from the authors of the legislation.

The community of Edison not only deserves this type of examination on our part, but they need it. The country needs this examination, too. This near tragedy should be a national, not just a local, wake-up call. We do not want to have to wait for 50 accidents in 50 States before we all benefit from the lessons learned as a result of this accident.

[Testimony resumes on p. 32.]

[The text of H.R. 4394 and H.R. 4616 follows:]

I

103D CONGRESS
2D SESSION

H. R. 4394

To provide for the establishment of mandatory State-operated comprehensive one-call systems to protect natural gas and hazardous liquid pipelines and all other underground facilities from being damaged by any excavations, and for other purposes.

IN THE HOUSE OF REPRESENTATIVES

MAY 11, 1994

Mr. PALLONE introduced the following bill; which was referred jointly to the Committees on Public Works and Transportation and Energy and Commerce

---

A BILL

To provide for the establishment of mandatory State-operated comprehensive one-call systems to protect natural gas and hazardous liquid pipelines and all other underground facilities from being damaged by any excavations, and for other purposes.

Be it enacted by the Senate and House of Representatives of the United States of America in Congress assembled,

SECTION 1. SHORT TITLE.

This Act may be cited as the "Comprehensive One-Call Notification Act of 1994".

SEC. 2. FINDINGS.

The Congress finds that—

(1) since the 1950s, steadily increasing development of infrastructure has resulted in the construction of underground facilities throughout the United States, including natural gas pipelines, hazardous liquid pipelines, water pipelines, steam pipelines, telephone lines, electric lines, fiber optic lines, cable television lines, sewer pipelines, and dedicated traffic control, emergency communication, and alarm lines;

(2) these underground facilities offer a safe and economical means of providing essential services to the public;

(3) the vast majority of accidents involving these facilities are caused by nearby excavation, demolition, or tunneling activities, known as third-party damage;

(4) accidents resulting from third-party damage present an unnecessary risk to public safety and the environment;

(5) costs arising from third-party damage are ultimately paid by consumers;

(6) in the case of interstate facilities, consumers in one State may pay for damages incurred in another State;

(7) to prevent third-party damage, the owners of some underground facilities have initiated one-call

(or "call before you dig") programs, and some States have mandated one-call programs, although the scope and effectiveness of these programs is inconsistent;

(8) to maximize the effectiveness of one-call programs, national standards are needed;

(9) these standards should apply, without exception, to all excavation near any underground facilities; and

(10) these standards should produce one-call systems which are simple to use, with a single telephone number established which excavators must call to obtain information on the location of any type of underground facility anywhere in the United States.

SEC. 3. DEFINITIONS.

For purposes of this Act, the term—

(1) "damage" means any impact or contact with an underground facility, its appurtenances, or its protective coating, or weakening of the support for the facility or protective housing, which requires repair;

(2) "excavation" means any operation in which earth, rock, or other material in the ground is moved, removed, or otherwise displaced by means of

any tools, equipment, or explosive, and includes, without limitation, grading, boring, milling, trenching, tunneling, scraping, tree and root removal, cable or pipe plowing, pile driving, wrecking, razing, rending, or removing any structure or mass material, but shall not include the tilling of soil for agricultural purposes to a depth of 18 inches or less;

(3) "facility operator" means any person who owns or operates an underground facility, except for any person who is the owner of real property wherein are located underground facilities for the purpose of furnishing services or materials only to himself or occupants of such property;

(4) "Secretary" means the Secretary of Transportation; and

(5) "underground facility" means any underground line, system, or structure used for producing, gathering, storing, conveying, transmitting, or distributing natural gas, hazardous liquids, communication, electricity, petroleum, petroleum products, water, steam, sewerage, or any other commodities the Secretary determines to be similar and appropriate.

SEC. 4. NATIONWIDE DEDICATED NUMBER.

Within 1 year after the date of enactment of this Act, the Federal Communications Commission shall establish a nationwide dedicated telephone number to be used by local or regional underground facility location services, and by one-call systems established pursuant to this Act.

SEC. 5. ESTABLISHMENT OF STATE ONE-CALL SYSTEMS.

(a) REQUIREMENT.—Each State shall, within 3 years after the date of enactment of this Act, establish a comprehensive statewide one-call notification system, in accordance with this Act, to protect all underground facilities from damage due to any excavation.

(b) STATE SANCTIONS FOR NONPARTICIPATION.—The Secretary may impose a prohibition, applicable to a State that does not comply with subsection (a), on the approval of any projects or the awarding of any grants under title 23, United States Code, other than projects or grants for safety where the Secretary determines, based on accident or other appropriate data submitted by the State, that the principal purpose of the project is an improvement in safety to resolve a demonstrated safety problem and likely will result in a significant reduction in, or avoidance of, accidents.

SEC. 6. ELEMENTS OF SYSTEM.

Each State one-call system established under section 5(a) shall—

(1) have a designated system operator;

(2) operate in all areas of the State containing underground facilities;

(3) apply to all excavations and to all underground facility operators, except as provided by this Act;

(4) employ mechanisms, such as the issuance of excavation or building permits, to ensure that the general public, and in particular all excavators, are aware of the one-call telephone number and the requirements and penalties of the State system relating to excavations;

(5) require that any person conducting an excavation must contact the one-call system at least 3 business days, and not more than 10 business days, before excavation begins;

(6) receive and record appropriate information from excavators about intended excavations, including—

(A) the name of the person contacting the one-call system;

(B) the name, address, and telephone number of the excavator; and

(C) the specific location of the intended excavation, along with the starting date thereof and a description of the intended excavation activity;

(7) inform excavators of the identity of facility operators who will be notified of the intended excavation;

(8) inform excavators of any procedures that the State has determined must be followed when excavating;

(9) inform facility operators of any intended excavations that may be in the vicinity of their underground facilities;

(10) require facility operators to locate and mark, in accordance with standards established by the State, their underground facilities in the vicinity of an intended excavation within no more than 3 business days after notification of such intended excavation, and to supervise such excavation as necessary;

(11) provide for penalties and enforcement as described in section 7;

(12) maintain records on each notice of intent to excavate for at least 7 years;

(13) establish procedures to promote the timely acquisition of information on previously unknown underground facility locations;

(14) provide for an appropriate waiver of timely compliance with system requirements in emergency circumstances in which public safety is endangered, as long as the one-call system is notified at the earliest practicable time;

(15) establish an appropriate schedule of fees to be imposed on facility operators to cover the costs of establishing, maintaining, and operating the one-call system; and

(16) provide an opportunity for citizen suits to enforce the requirements of this section.

SEC. 7. PENALTIES AND ENFORCEMENT.

(a) GENERAL PENALTIES.—Each State one-call system established under section 5(a) shall provide that any excavator or facility operator who violates the requirements of the system shall be liable for a civil penalty of not more than $25,000 for each violation for each day that violation persists, except that the maximum civil penalty shall not exceed $500,000 for any related series of violations and the minimum civil penalty for a violation shall be not less than $250.

(b) INCREASED PENALTIES.—If a violation results in damage to an underground facility resulting in death, serious bodily harm, or actual damage to property exceeding $50,000, or damage to an underground hazardous liquid pipeline facility resulting in the release of more than 50 barrels of product, the penalties may be increased, and an additional penalty of imprisonment may be assessed.

(c) DECREASED PENALTIES.—A State one-call system may provide for reduced penalties for a violation, that results in or could result in damage, that is promptly reported by the violator.

(d) INJUNCTIVE RELIEF.—Each State one-call system shall provide for appropriate injunctive relief.

(e) REVOCATION OF LICENSE.—Each State one-call system shall include procedures for the revocation of a license or permit to do business of any excavator determined to be a habitual violator of the requirements of the system.

(f) IMMEDIATE CITATION OF VIOLATIONS.—A State one-call system may include procedures for issuing a citation of violation at the site and time of the violation.

SEC. 8. DEPARTMENT OF TRANSPORTATION.

(a) COORDINATION WITH OTHER RESPONSIBILITIES.—The Secretary shall coordinate the implementation of this Act with the implementation of the Natural Gas Pipeline Safety Act of 1968 and the Hazardous Liquid Pipeline Safety Act of 1979.

(b) MODEL PROGRAM.—Within 1 year after the date of enactment of this Act, the Office of Pipeline Safety of the Department of Transportation shall draft and make available to States a model one-call system program, along with such additional guidance as the Secretary considers appropriate, to assist the States in complying with this Act. Such model program may be amended in response to reports submitted by the States pursuant to section 10.

(c) PUBLIC EDUCATION.—The Secretary shall develop public service announcements to be broadcast or published to educate the public about one-call notification systems, including the national phone number.

SEC. 9. ALTERNATE FORM OF SYSTEM.

A State that wishes to establish or maintain a one-call system that differs from the requirements of this Act may petition the Secretary for approval of such system. The Secretary shall approve such a petition within 1 month after receipt of the petition, if the proposed system is at least as protective of the public health and safety as a system described in this Act. The Secretary shall inform a State of the reasons for the denial of a petition under this section.

SEC. 10. STATE REPORTS.

Within 54 months after the date of enactment of this Act, and annually thereafter each State shall report to Congress and the Office of Pipeline Safety of the Department of Transportation on the status of their one-call notification system and its requirements. The report shall contain data on the operation and effectiveness of the one-call system including—

(1) the status of its law establishing the one-call system;

(2) the number of notification requests received annually;

(3) the effectiveness of the method of underground facility marking required;

(4) the degree of excavator compliance;

(5) the number of incidents where underground facilities were damaged and the type of damage to such facilities;

(6) the number of deaths and injuries and the estimate amount of property loss resulting from damage to underground facilities;

(7) the extent to which all underground facilities participate; and

(8) any other relevant information.

○

I

103D CONGRESS
2D SESSION

H. R. 4616

To amend the Natural Gas Pipeline Safety Act of 1968 and the Hazardous Liquid Pipeline Safety Act of 1979 to improve natural gas and hazardous liquid pipeline safety, in response to the natural gas pipeline accident in Edison, New Jersey, and for other purposes.

IN THE HOUSE OF REPRESENTATIVES

JUNE 21, 1994

Mr. PALLONE introduced the following bill; which was referred jointly to the Committees on Energy and Commerce and Public Works and Transportation

---

A BILL

To amend the Natural Gas Pipeline Safety Act of 1968 and the Hazardous Liquid Pipeline Safety Act of 1979 to improve natural gas and hazardous liquid pipeline safety, in response to the natural gas pipeline accident in Edison, New Jersey, and for other purposes.

Be it enacted by the Senate and House of Representatives of the United States of America in Congress assembled,

SECTION 1. SHORT TITLE.

This Act may be cited as the "Natural Gas Pipeline Safety Improvement Act of 1994".

SEC. 2. RECOVERY BY SECRETARY OF TRANSPORTATION OF COSTS OF INVESTIGATION OF CERTAIN PIPELINE ACCIDENTS.

(a) NATURAL GAS PIPELINE ACCIDENTS.—Section 14 of the Natural Gas Pipeline Safety Act of 1968 (49 U.S.C. App. 1681) is amended by adding at the end the following:

"(g)(1)(A) Subject to paragraphs (2) and (3), the Secretary may recover from any person who engages in the transportation of gas, or who owns or operates pipeline facilities, the costs incurred by the Secretary—

"(i) in investigating an accident with respect to such transportation or facilities; and

"(ii) in overseeing the response of the person to the accident.

"(B) For the purposes of this paragraph, the costs incurred by the Secretary in an investigation of an accident may include the cost of hiring additional personnel (including personnel to support monitoring activities by the Office of Pipeline Safety), the cost of tests or studies, and travel and administrative costs associated with the investigation.

"(2) The Secretary may not recover costs under this subsection with respect to an accident unless the accident—

"(A) results in death or personal injury; or

"(B) results in property damage (including the cost of any lost natural gas) and environmental damage (including the cost of any environmental remediation) in an amount in excess of $250,000.

"(3) The amount that the Secretary may recover under this subsection with respect to an accident may not exceed $500,000.

"(4)(A) Amounts recovered by the Secretary under this subsection shall be available to the Secretary for purposes of the payment of the costs of investigating and overseeing responses to accidents under this subsection. Such funds shall be available to the Secretary for such purposes without fiscal year limitation.

"(B) Such amounts shall be used to supplement and not to supplant other funds made available to the Secretary for such purposes.".

(b) HAZARDOUS LIQUID PIPELINE ACCIDENTS.—Section 211 of the Hazardous Liquid Pipeline Safety Act of 1979 (title II of Public Law 96–129; 49 U.S.C. App. 2010) is amended by adding at the end the following:

"(g)(1)(A) Subject to paragraphs (2) and (3), the Secretary may recover from any person who engages in the transportation of hazardous liquids, or who owns or operates pipeline facilities, the costs incurred by the Secretary—

"(i) in investigating an accident with respect to such transportation or facilities; and

"(ii) in overseeing the response of the person to the accident.

"(B) For the purposes of this paragraph, the costs incurred by the Secretary in an investigation of an accident may include the cost of hiring additional personnel (including personnel to support monitoring activities by the Office of Pipeline Safety), the cost of tests or studies, and travel and administrative costs associated with the investigation.

"(2) The Secretary may not recover costs under this subsection with respect to an accident unless the accident—

"(A) results in death or personal injury; or

"(B) results in property damage (including the cost of any lost hazardous liquid) and environmental damage (including the cost of any environmental remediation) in an amount in excess of $250,000.

"(3) The amount that the Secretary may recover under this subsection with respect to an accident may not exceed $500,000.

"(4)(A) Amounts recovered by the Secretary under this subsection shall be available to the Secretary for purposes of the payment of the costs of investigating and overseeing responses to accidents under this subsection. Such funds shall be available to the Secretary for such purposes without fiscal year limitation.

"(B) Such amounts shall be used to supplement and not to supplant other funds made available to the Secretary for such purposes.".

SEC. 3. GRANTS TO STATES AND ONE-CALL NOTIFICATION SYSTEMS TO PROMOTE USE OF SUCH SYSTEMS.

(a) GRANTS TO STATES.—Subsection (c) of section 20 of the Natural Gas Pipeline Safety Act of 1968 (49 U.S.C. App. 1687) is amended by adding at the end the following: "The Secretary may make a grant to a State for development and establishment of a one-call notification system only if the State ensures that the cost of establishing and operating the system are shared equitably by persons owning or operating underground facilities.".

(b) GRANTS TO SYSTEMS.—Such subsection is further amended—

(1) by striking "GRANTS TO STATES.—" and inserting "GRANTS TO STATES AND SYSTEMS.—(1)"; and

(2) by adding at the end the following:

"(2)(A) The Secretary may also make grants to one-call notification systems for activities relating to the promotion of the utilization of such systems.

"(B) The Secretary shall ensure that the Federal share of the cost of the activities referred to in subparagraph (A) under any grant made under this paragraph does not exceed 50 percent of the cost of such activities.".

(c) SANCTIONS.—Subsection (b)(9) of such section is amended by inserting ", or that would provide for effective civil or criminal penalty sanctions or equitable relief appropriate to the nature of the offense" after "12 of this Act".

(d) CONFORMING AMENDMENT.—Subsection (f) of such section is amended by striking out "subsection (c)" and inserting in lieu thereof "subsection (c)(1)".

SEC. 4. PREVENTION OF DAMAGE TO PIPELINE FACILITIES.

(a) NATURAL GAS PIPELINE FACILITIES.—Section 14(a) of the Natural Gas Pipeline Safety Act of 1968 (49 U.S.C. App. 1681(a)) is amended by inserting after "and training activities" the following: "and promotional activities relating to prevention of damage to pipeline facilities".

(b) HAZARDOUS LIQUID PIPELINE FACILITIES.—Section 211(a) of the Hazardous Liquid Pipeline Safety Act of 1979 (title II of Public Law 96–129; 49 U.S.C. App. 2010(a)) is amended by inserting after "and training activities" the following: "and promotional activities relating to prevention of damage to pipeline facilities".

SEC. 5. ELECTRONIC DATA ON PIPELINE FACILITIES FOR RISK ASSESSMENT AND SAFETY PLANNING.

(a) AUTHORITY TO DEVELOP.—The Secretary of Transportation may develop an electronic data base containing uniform information on the nature, extent, and geographic location of pipeline facilities. The purpose of the data base shall be to provide information on such facilities to the Secretary, owners of pipeline facilities, as persons engaged in transporting gas or hazardous liquids through pipeline facilities, and for secured use by State agencies concerned with land use planning, environmental regulation, and pipeline regulatory oversight, in order to facilitate risk assessment and safety planning with respect to such facilities.

(b) CONTRACT AND GRANT AUTHORITY.—(1) Subject to paragraph (2), the Secretary may develop the data base described under subsection (a) by entering into contracts or cooperative agreements with any entity that the Secretary determines appropriate for that purpose and by making grants to States or institutions of higher education for that purpose.

(2) The Secretary shall ensure that the Federal share of the cost of any activities carried out under a grant or

cooperative agreement made under this subsection does not exceed 50 percent of the cost of such activities.

(c) USE OF GEOGRAPHIC INFORMATION SYSTEM TECHNOLOGY.—In developing the data base described in subsection (a), the Secretary shall, to the maximum extent practicable, develop a data base that—

(1) utilizes Geographic Information System technology or any similar technology providing data of an equivalent quality and usefulness; and

(2) permits ready incorporation of data and information from a variety of sources.

(d) DEFINITION.—For purposes of this section, the term "pipeline facility" has the meaning given such term in section 20(e) of the Natural Gas Pipeline Safety Act of 1968 (49 U.S.C. App. 1687(e)).

SEC. 6. AUTHORIZATION OF APPROPRIATIONS.

(a) NATURAL GAS PIPELINE SAFETY ACT OF 1968.—(1) Section 17(a) of the Natural Gas Pipeline Safety Act of 1968 (49 U.S.C. App. 1684(a)) is amended—

(A) in paragraph (12), by striking "and";

(B) by striking paragraph (13); and

(C) by adding after paragraph (12) the following new paragraphs:

"(13) $20,000,000 for the fiscal year ending September 30, 1995;

"(14) $30,000,000 for the fiscal year ending September 30, 1996; and

"(15) $35,000,000 for the fiscal year ending September 30, 1997.".

(2) Section 17(c) of the Natural Gas Pipeline Safety Act of 1968 (49 U.S.C. App. 1684(c)) is amended by striking "and $10,000,000 for the fiscal year ending September 30, 1995" and inserting in lieu thereof "$16,500,000 for the fiscal year ending September 30, 1995, $19,000,000 for the fiscal year ending September 30, 1996, and $21,500,000 for the fiscal year ending September 30, 1997".

(b) HAZARDOUS LIQUID PIPELINE SAFETY ACT OF 1979.—Section 214(a) of the Hazardous Liquid Pipeline Safety Act of 1979 (49 U.S.C. App. 2013(a)) is amended—

(1) in paragraph (12), by striking "and";

(2) by striking paragraph (13); and

(3) by adding after paragraph (12) the following new paragraphs:

"(13) $7,000,000 for the fiscal year ending September 30, 1995;

"(14) $10,000,000 for the fiscal year ending September 30, 1996; and

"(15) $11,000,000 for the fiscal year ending September 30, 1997.".

SEC. 7. SITING OF INTERSTATE TRANSMISSION FACILITIES.

(a) SITING GUIDELINES.—Within 2 years after the date of enactment of this Act, the Federal Energy Regulatory Commission shall review its practices and guidelines for siting natural gas interstate transmission facilities in urban areas to determine whether changes are needed in the areas of—

(1) selecting routes for pipelines; and

(2) determining the appropriate width of rights-of-way.

(b) EDUCATIONAL INFORMATION FOR LOCAL JURISDICTIONS.—(1)(A) Within 2 years after the date of enactment of this Act, the Secretary, in consultation with the Federal Energy Regulatory Commission, shall make educational information available, regarding natural gas interstate transmission facilities permits and rights-of-way and issues with respect to development in the vicinity of such interstate transmission facilities, for distribution to appropriate agencies of local governments with jurisdiction over the lands through which natural gas interstate transmission facilities pass.

(B) For purposes of this section, the term "interstate transmission facilities" has the meaning given such term in section 2(8) of the Natural Gas Pipeline Safety Act of 1968 (49 U.S.C. App. 1671(8)).

(2)(A) Within 2 years after the date of enactment of this Act, the Secretary shall make educational information available, regarding hazardous liquid interstate pipeline facilities rights-of-way and issues with respect to development in the vicinity of such interstate pipeline facilities, for distribution to appropriate agencies of local governments with jurisdiction over the lands through which hazardous liquid interstate pipeline facilities pass.

(B) For purposes of this paragraph, the term "interstate pipeline facilities" has the meaning given such term in section 202(5) of the Hazardous Liquid Pipeline Safety Act of 1979 (49 U.S.C. App. 2001(5)).

(3) There are authorized to be appropriated to the Secretary of Energy for carrying out this subsection, $2,000,000, to remain available until expended.

SEC. 8. DUMPING WITHIN PIPELINE RIGHTS-OF-WAY.

(a) NATURAL GAS PIPELINE SAFETY ACT OF 1968.—

(1) AMENDMENT.—The Natural Gas Pipeline Safety Act of 1968 (49 U.S.C. App. 1671 et seq.)

is amended by adding at the end the following new section:

"SEC. 22. DUMPING WITHIN PIPELINE RIGHTS-OF-WAY.

"(a) PROHIBITION.—No person shall excavate within the right-of-way of a natural gas interstate transmission facility, or any other limited area in the vicinity of such interstate transmission facility established by the Secretary, and dispose solid waste therein.

"(b) DEFINITION.—For purposes of this section, the term 'solid waste' has the meaning given such term in section 1004(27) of the Solid Waste Disposal Act (42 U.S.C. 6903(27)).".

(2) CONFORMING AMENDMENTS.—Section 11(a)(1) and (c)(1) of the Natural Gas Pipeline Safety Act of 1968 (49 U.S.C. App. 1679a(a)(1) and (c)(1)) is amended by striking "or section 20(h)" and inserting in lieu thereof ", section 20(h), or section 22(a)".

(b) HAZARDOUS LIQUID PIPELINE SAFETY ACT OF 1979.—

(1) AMENDMENT.—The Hazardous Liquid Pipeline Safety Act of 1979 (49 U.S.C. App. 2001 et seq.) is amended by adding at the end the following new section:

"SEC. 221. DUMPING WITHIN PIPELINE RIGHTS-OF-WAY.

"(a) PROHIBITION.—No person shall excavate within the right-of-way of a hazardous liquid interstate pipeline facility, or any other limited area in the vicinity of such interstate pipeline facility established by the Secretary, and dispose solid waste therein.

"(b) DEFINITION.—For purposes of this section, the term 'solid waste' has the meaning given such term in section 1004(27) of the Solid Waste Disposal Act (42 U.S.C. 6903(27)).".

(2) CONFORMING AMENDMENTS.—Section 208(a)(1) and (c)(1) of the Hazardous Liquid Pipeline Safety Act of 1979 (49 U.S.C. App. 2007(a)(1) and (c)(1)) is amended by inserting "or section 221(a)" after "section 207(a)".

SEC. 9. PERIODIC INSPECTION BY INSTRUMENTED INTERNAL INSPECTION DEVICES.

(a) NATURAL GAS PIPELINE SAFETY ACT OF 1968.—Section 3(g)(2) of the Natural Gas Pipeline Safety Act of 1968 (49 U.S.C. App. 1672(g)(2)) is amended—

(1) by striking "Not later than 3 years after the date of the enactment of this paragraph" and inserting in lieu thereof "Not later than 1 year after the date of the enactment of the Natural Gas Pipeline Safety Improvement Act of 1994"; and

(2) in the first sentence, by inserting "and shall prescribe a schedule or schedules for such inspections" after "operator of the pipeline".

(b) HAZARDOUS LIQUID PIPELINE SAFETY ACT OF 1979.—Section 203(k)(2) of the Hazardous Liquid Pipeline Safety Act of 1979 (49 U.S.C. App. 2002(k)(2)) is amended—

(1) by striking "Not later than 3 years after the date of the enactment of this paragraph" and inserting in lieu thereof "Not later than 1 year after the date of the enactment of the Natural Gas Pipeline Safety Improvement Act of 1994"; and

(2) in the first sentence, by inserting "and shall prescribe a schedule or schedules for such inspections" after "operator of the pipeline".

SEC. 10. PROMOTING PUBLIC AWARENESS FOR NEIGHBORS OF PIPELINES.

(a) NATURAL GAS PIPELINE SAFETY ACT OF 1968.—Section 18 of the Natural Gas Pipeline Safety Act of 1968 (49 U.S.C. App. 1685) is amended by adding at the end the following new subsections:

"(c) PROMOTING PUBLIC AWARENESS FOR NEIGHBORS OF PIPELINES.—Not later than 1 year after the date of enactment of this subsection, and annually thereafter, the owner or operator of each interstate transmission facility shall notify all residents within 1000 yards, or such other distance as the Secretary determines appropriate, of such interstate transmission facility of—

"(1) the general location of the interstate transmission facility;

"(2) a request for reporting of any instances of excavation or dumping on or near the interstate transmission facility;

"(3) a phone number to use to make such reports; and

"(4) appropriate procedures for such residents to follow in response to accidents concerning interstate transmission facilities.

"(d) PUBLIC EDUCATION.—The Secretary shall develop, in conjunction with appropriate representatives of the natural gas pipeline industry, public service announcements to be broadcast or published to educate the public about pipeline safety.".

(b) HAZARDOUS LIQUID PIPELINE SAFETY ACT OF 1979.—Section 212 of the Hazardous Liquid Pipeline Safety Act of 1979 (49 U.S.C. App. 2011) is amended by adding at the end the following new subsections:

"(e) PROMOTING PUBLIC AWARENESS FOR NEIGHBORS OF PIPELINES.—Not later than 1 year after the date of enactment of this subsection, and annually thereafter, the owner or operator of each interstate pipeline facility shall notify all residents within 1000 yards, or such other distance as the Secretary determines appropriate, of such interstate pipeline facility of—

"(1) the general location of the interstate pipeline facility;

"(2) a request for reporting of any instances of excavation or dumping on or near the interstate pipeline facility;

"(3) a phone number to use to make such reports; and

"(4) appropriate procedures for such residents to follow in response to accidents concerning interstate pipeline facilities.

"(f) PUBLIC EDUCATION.—The Secretary shall develop, in conjunction with appropriate representatives of the hazardous liquid pipeline industry, public service announcements to be broadcast or published to educate the public about pipeline safety.".

SEC. 11. REMOTELY OR AUTOMATICALLY CONTROLLED VALVES.

Section 3 of the Natural Gas Pipeline Safety Act of 1968 (49 U.S.C. App. 1672) is amended by adding at the end the following new subsection:

84–088 O - 94 - 2

"(l) REMOTELY OR AUTOMATICALLY CONTROLLED VALVES.—Not later than 18 months after the date of enactment of this subsection, the Secretary shall issue regulations requiring the installation and use, wherever technically and economically feasible, of remotely or automatically controlled valves that are reliable and capable of shutting off the flow of gas in the event of an accident, including accidents in which there is a loss of the primary power source. In developing proposed regulations, the Secretary shall consult with, and give special consideration to recommendations of, appropriate groups from the gas pipeline industry, such as the Gas Research Institute.".

SEC. 12. BASELINE INFORMATION.

(a) NATURAL GAS PIPELINE SAFETY ACT OF 1968.—Section 3(g) of the Natural Gas Pipeline Safety Act of 1968 (49 U.S.C. App. 1672(g)) is amended by adding at the end the following new paragraph:

"(3) BASELINE INFORMATION.—Before transporting natural gas through a pipeline which, because of its design, construction, or replacement, is required by regulations issued under paragraph (1) to accommodate the passage of instrumented internal inspection devices, the owner or operator of such pipeline shall, using such a device, obtain baseline information with respect to the safety of the pipeline.''.

(b) HAZARDOUS LIQUID PIPELINE SAFETY ACT OF 1979.—Section 203(k) of the Hazardous Liquid Pipeline Safety Act of 1979 (49 U.S.C. App. 2002(k)) is amended by adding at the end the following new paragraph:

"(3) BASELINE INFORMATION.—Before transporting hazardous liquids through a pipeline which, because of its design, construction, or replacement, is required by regulations issued under paragraph (1) to accommodate the passage of instrumented internal inspection devices, the owner or operator of such pipeline shall, using such a device, obtain baseline information with respect to the safety of the pipeline.''.

○

Mr. SHARP. With that, I would like to welcome our first panel. As I understand, the two Senators may join us at a later point due to Senate activity, so we will work them in as the opportunity arises during our hearing. First, I recognize the distinguished gentleman from Illinois.

Mr. HASTERT. Well, I thank the chairman, I want to commend you, Mr. Chairman, for holding this timely hearing on pipeline safety and on one-call notification initiatives. I know that I speak for all the members of the subcommittee in saying that although the natural gas pipeline industry has a good safety record, it is the responsibility of both the industry and Congress to strive for improved safety standards. This hearing will afford us the opportunity to discuss various safety initiatives that may warrant implementation next year when the Natural Gas Pipeline Safety Act is up for reauthorization.

I am particularly pleased that we will be discussing the proposal introduced by Congressman Pallone for a comprehensive national one-call notification system. I agree with the goals of such a system and hope with your assistance, Mr. Chairman, we can resolve any disagreements we may have about the mechanics of this legislation and pass a responsible one-call notification system yet this year.

Finally, Mr. Chairman, I have an article entitled "Pipeline Integrity Program", which I would like to submit for the record at the appropriate time. And also from my other tour of duty in this Congress on Public Works and Transportation, we looked at the pipeline issue to some extent and looking at it and going out and looking at pipeline operations, I have come to the conclusion that there is a lot of product that is moved in this country, natural gas, gasoline and other petroleum products, crude, that is transported very, very safely in this country, much more safely than it would be transported on the surface, either in rail cars or on trucks, and we don't want to do anything to impede this industry.

We just want to make it safer and have also known the great amount of research that they have done with smart pigs and being able to detect from the inside out how safe these lines are. So we want to encourage that type of operation and the good work that industry has done. But on the other hand, we know that from the outside in, construction work and those types of things are usually the causes for the breaches of these lines and putting them in a dangerous situation, so we want to make sure from our point of view that we can do a better job.

I know in my own State we have the Julie program for when you call State-wide to do any digging. It has been pretty successful and I would hope that we would not impose lesser restrictions than we have already in place in the States with Representative Pallone's legislation. We would like to work it out and work with him on that.

So I would like to thank you and also ask that Congressman Moorhead's opening statement be put in the record.

Mr. SHARP. Without objection, so ordered.

[The opening statement of Mr. Moorhead and the article referred to follow:]

STATEMENT OF HON. CARLOS J. MOORHEAD

Mr. Chairman, I am pleased that we are holding this hearing on H.R. 4394, The Comprehensive One-call Notification Act of 1994 and other pipeline safety initiatives. This hearing is extremely timely in light of the recent natural gas pipeline accidents.

The recent pipeline accidents have emphasized what we already knew: that we must make every effort to make pipelines as safe as possible. Since nearly ⅔ of all damage to pipelines is the result of third-party activity, one of the most effective and low cost mechanisms for improving the safety of pipelines is to establish mandatory one-call programs. My home State of California has had very successful mandatory one-call programs for a number of years.

Thus, I support the goals of the bill introduced by Mr. Pallone to establish Statewide, mandatory one-call systems. However, I do have some questions about the specifics of his bill. Of most notable concern are the severe penalties excavators and others are subject to for failure to usethe one-call system. I do not believe failure to call a one-call system, without some element of willfulness, should be a criminal act. In addition, I think stiff penalties could in fact have a negative effect on encouraging people to report minor damage to pipelines.

I also have concerns about whether the provisions in H.R. 4616 are cost-effective improvements of pipeline safety. In addition, I am concerned that the $35 million budget request from the Department of Transportation for the pipeline safety program is reasonable. Since the pipeline safety program is funded 100 percent by user fees I hope that the Department does not see this as an unlimited source of funds which can be tapped at will, especially since pipeline ratepayers ultimately bear the cost of these fees. I hope the Department can show today why it will require nearly $100 per mile to regulate the safety of natural gas and hazardous liquid pipelines.

Again, while I have questions about the effectiveness of the pipeline safety proposals, I am strongly in favor of a onecall initiative which will help reduce the risk to human health and the environment and which will save underground facility owners millions of dollars in costly repairs from third-party damage to such facilities. I look forward to working with you this year to move this important piece of legislation.

Thank you, Mr. Chairman.

RISK ASSESSMENT

Pipeline Integrity Program

(Above) Close interval surveys supply a detailed description of potentials and a good indicator of where significant coating deterioration has occured. Results help improve the accuracy of the cathodic protection rating in the risk assessment program.

(Above Right) Pipeline integrity projects ranked by the program include everything from coating repairs to replacement of existing stream crossings.

by Philip J. Dusek, Chief Engineer/Corrosion Control, Natural Gas Pipeline Co. of America, Lombard, Ill.

A number of forces are exerting pressure on natural gas pipeline operators to develop integrity programs for their systems.

Regulators are pushing for a risk-free society, while deregulation is prompting company reorganizations that often result in deep work-force reductions. These contrasting forces require pipeline operators to rethink system integrity programs to optimize capital and personnel resources.

Many new methods are being used today to meet the challenge, most of which take an active approach toward improving the pipeline's integrity.

Today, the term "pipeline integrity" is being applied to everything from smart pigging to risk assessment programs. A true integrity program would include these items, but also many other operational issues intended to protect the pipeline from damage by outside forces.

Natural Gas Pipeline Co. of America has developed an integrity program.

NGPL operates approximately 13,000 miles of large-diameter parallel gas pipelines, which extend from traditional supply areas to the Chicago area. Line Number 1, the 24-in. Amarillo-to-Chicago mainline, was built in 1931, and parts of it are still in operation today. More than 85% of the NGPL system is more than 25 years old, and continues to provide very reliable service.

The company operated for many years with specialized crews dedicated to pipeline systems, and a corrosion department. Under this organization, employees developed an intimate knowledge of the pipeline and related integrity issues. NGPL relied on this knowledge to develop its integrity program.

Emphasis shifted to a decentralized organization in 1986 due to an early retirement program and reorganization. The new organization did not include a corrosion department nor crews dedicated to pipeline work.

All NGPL field facilities are now operated by a single department. The system is divided into five operations areas, with numerous districts reporting to the area directors. Individual districts are responsible for all technical and integrity issues within their districts, supported technically by service engineers located throughout the system. Districts are responsible for generating all integrity work-order requests.

A pipeline integrity program was first initiated with the area personnel submitting a multi-year plan of integrity projects to management for budgeting. The plans were very good, but upper management found that there was no way of judging which programs were most deserving of budget dollars without an overall program.

This was the basis for the establishment of the current pipeline integrity task group. The group's objective was to develop an overall strategy, and methodically evaluate and identify

Helps Optimize Resources

integrity projects.

The pipeline integrity group, formed in 1991, consists of personnel from operations and engineering divisions and covers many technical areas. The initial assignment was to develop a rating method for projects that are submitted to management, and also recognize areas needing remedial work that may not be submitted by the areas.

The integrity group first had to define the factors to be considered, and how to accumulate and organize data. Pipeline information was not centralized, and compilation of data appeared to be an imposing task.

Another problem was how to rate the importance of each pipeline factor. The ideal situation would be a statistical comparison of problems compared to failures and consequences.

The pipeline integrity group decided on an outside sorting and gathering system, and considered geographical information systems, decision tree analysis, risk assessment algorithms and miscellaneous ranking systems.

A risk assessment algorithm was chosen as the basis for managing the rating process. Numerous algorithms are available, and NGPL selected one developed by the Pipeline Research Committee of the American Gas Association. The selection was based on the PRC algorithm's comprehensive inclusion of many integrity-related factors. The group also felt the outside program would be more objective and better accepted by regulators and company personnel.

After the algorithm was chosen, a method had to selected for collecting and compiling data. The obvious choice was one of the most important decisions in implementing a program, so someone knowledgeable about data-base software should be included in the process. Questions should be asked before the software is chosen, including:

■ Who will use the program? A mainframe or data-base program will allow maximum flexibility and manipulation of data in a centralized organization, but in a decentralized organization there will be many users. A spreadsheet program allows more universal use of program between field and office.

■ What type of software is available in the system?

■ Is there a GIS system, or will one be implemented? If so, the algorithm can be tied into the system with very little effort and yield excellent results.

A spreadsheet program was selected for data compilation. Data entries were kept separate from algorithm calculations to allow for changes and different queries of the information. The independent data entries used actual dates, test pressures and other raw entries so the program could be used as a data base. Independent data also permit changing criteria in the algorithm and aging the pipeline automatically in the program without re-entering information every year.

The integrity group chose a risk assessment program to determine

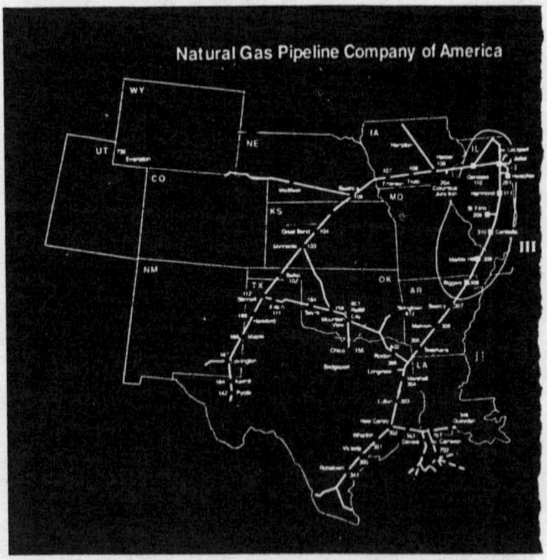

Natural Gas Pipeline Co. of America operates approximately 13,000 miles of large-diameter natural gas transmission pipelines, which extend from supply sources to the Chicago area.

rank. Numerous programs are available to pipeline operators.

These programs consider the probability and consequences of failure to determine a risk-assessment ranking number. Probability factors are generally empirical, based on industry experience with a mix of absolute and subjective issues. Consequence factors are obvious, but the ratings applied are random and dependent on the extent of failure mode.

A risk assessment program is not an exact analysis that either qualifies or proves the safety of any particular section due to a ranking number. The number is only a comparison.

The relative risk is the probability of failure multiplied by the consequence of failure. Probability factors include coating type, cathodic protection, stress corrosion cracking, pipe condition, coating condition, wall thickness, soil type, leak history and many other items. Consequence factors include class location and propagation of ductile and brittle fractures.

Each factor is assigned a weighted number defined in the algorithm instructions to correspond with a level of probability or risk associated with the condition described by each. This number is then inserted into the algorithm to calculate the risk index. For example, a pipeline coated with coal-tar enamel will have a value of 2.0, but a bare line with a higher probability of failure will have a value of 20.

The algorithm equation is long and involved. It is contained in the PRC report, "Project PR3-919: Methods For Prioritizing Pipeline Maintenance and Rehabilitation," which can be obtained from the American Gas Association.

Many listed factors are measurable values, while others (such as pipe and coating condition) are subjective. Unfortunately, some of the more important factors are subjective, which can result in errors due to poor judgement or inconsistent evaluation. It is important that much of this type of information have a basis, such as buried pipeline inspection reports or similar examinations, to confirm judgements. The larger the sampling, the more accurate the judgements will be.

Whenever there were insufficient recent data available, the worst-case scenario was always assumed as a fail-safe mechanism. This usually made these sections appear worse than they actually were and provided a stimulus for local management to obtain the correct data.

No algorithm will suit the need of every pipeline company. NGPL chose one developed by the PRC and entered some items unique to the pipeline system. This data were added within the existing framework so any changes would only make the system more conservative.

Many factors in the algorithm have the same value assigned for numerous variations, such as the different types of long seams in line pipe. Although seamless, DSAW, ERW and flashweld seams had the same numerical rating, they were entered separately so that the rating could be changed for one or more factors later if necessary, and also use the program as a data base.

Some of the significant changes NGPL made to the algorithm include:

- Coatings not listed on coating selection sheet were included, and a rating factor assigned to each.
- A more favorable rating was given for close interval surveys. These data were more valuable than normal annual surveys taken at test lead stations. Date and type of survey performed are entered.
- Current density was added to the cathodic protection rating for future use.
- A time-variable ramping equation was added to the stress corrosion cracking factor in the algorithm. The equation increases the rating as the most recent hydrotest ages.
- A section on smart pigging was included, along with the date performed, to help determine pipe condition. The data will lose its accuracy as the smart pig information ages, and the index will go higher.

Data Collection

Although compilation of data in a decentralized system was a formidable task, it spurred one of the biggest—and unexpected—successes of the program.

Data entry was broken down into two phases, first the head office and then the field. Engineering data were collected at the Lombard headquarters from numerous departments holding many separate data bases. Task group members enlisted the help of the individual districts, which supplied much of the operational data. The members toured areas of the company to begin the data collection process and standardize data acquisition.

As the task group gathered the information, several benefits surfaced during data collection:

■ Reinforced the need for accurate and complete buried pipeline inspection reports.

■ Developed ownership in the pipeline integrity program.

■ Developed a higher level of interest in corrosion control program, to reduce index numbers.

■ Provided a communication link between field and head office for the integrity information before a work order is requested.

■ Forced a review and compilation of various data fields as data were collected.

The compilation of data for this program was an improvement over the centralized system, with different departments holding separate parts of the whole picture. Employees can now review all integrity issues on a single-sheet printout.

Section Size

The next hurdle for entering data was breaking down the pipeline sections. Setting section boundaries according to significant changes on the system proved to provide the most accurate rankings.

Due to the size of the NGPL transmission system and the fact it is basically located in rural areas, the existing 15-mile valve sections were used as the section limits. A few test cases were run where the pipeline conditions were known, and the results compared favorably.

After the data for the first year were compiled, it was soon apparent the 15-mile length was too long. A surprising number of changes, in class, coatings and other factors, were found to occur within a 15-mile section.

A fail-safe approach is always taken by entering the worst-case data for each section. If a valve section has a high population density area, low cathodic protection readings in another, and poor coatings in a third, the values were assumed for the entire section. This raised the index for many of the 15-mile sections to high but inaccurate numbers.

It was decided to break down the system into more discreet sections, as dictated by conditions on the pipeline. Sections were determined based on known engineering and design differences. This system data re-entry took almost a man-year to accomplish and increased the number of sections from 520 to 3,732. The average section's risk number decreased, from 73,496 to 54,163, and the length went from the original 15 miles to 2.24 miles.

The section size will also be determined by the number of factors covered in the algorithm. The more factors installed, the more natural section breaks will occur. The number of these sections can be controlled by ignoring some section breaks based on factors that don't significantly affect the index number.

System Use

NGPL's integrity program has proven useful for many integrity projects as well as everyday database requests, including:

■ Ranking integrity project requests for capital budgeting.

■ Identifying areas that need examination. For example, a high index number identified a pipeline with some previous SSC problems, and a field investigation turned up a badly corroded line where no work had been planned. The line was immediately scheduled for replacement.

■ Local district superintendents use risk index numbers to plan their integrity-related work, such as close interval surveys and recoating projects that will help reduce index numbers.

■ Comprehensive database has been developed that covers a wide field of pipeline items and is useful in developing listings for sections that have been smart pigged, where close-interval surveys have been performed and which sections can accommodate smart pigs.

A GIS system may be incorporated into the program to allow a direct correlation of smart pigging information and close interval surveys. Shorted casings may also be added to the program, along with internal corrosion factors.

Conclusions

The risk assessment program is a very valuable tool for identifying areas that may need remedial work. However, it is composed of many subjective evaluations and cannot predict failure nor ensure good performance.

The program is an excellent data management tool that enables a pipeline operator to combine all available information needed to make integrity decisions. The integrity of a pipeline is continually changing, and any program should be updated on a regular basis. **P&GJ**

Mr. SHARP. Does the gentleman from Idaho have an opening statement?

Mr. CRAPO. Just a very brief one. I, too, have dealt with this issue in the State-wide arena and am aware of the need for a State-wide system. I am looking for the explanation of why we need to federalize this approach and why it can't be left up to the States with some Federal support and I would just encourage the witnesses to express concern—or explain that kind of a need and why we do need to do this at the Federal level.

Mr. SHARP. The Chair thanks the gentleman.

I just, may, at the outset, articulate, I think we share a common goal. Many people believe the one-call system is the best possible technique that appears to be available to both protect workers who engage in excavation and to protect underground facilities as well as the risks to others living in the area of these facilities. The question in all of our minds is what is the best technique we can use at the Federal level to get the States—there are a number of States doing it and doing it successfully, but get all States to do it because the pipelines, of course, are interstate operations with an interstate safety program.

It starts with a Federal interest, but our goal is not to override State activities. Our goal is to encourage them to participate. This can be done with simply no requirements on the State treasury whatsoever. Indeed, I think many will testify today that programs are run successfully because they actually save money in the utility system that is protected, and because they more than finance themselves in the process.

[The prepared statement of Hon. Edward J. Markey follows:]

STATEMENT OF HON. EDWARD J. MARKEY

Thank you, Mr. Chairman. As a member of this subcommittee and as chairman of the Subcommittee on Telecommunications and Finance, I have a particular interest in the integrity, security and reliability of our underground facilities. I applaud your efforts, Mr. Chairman, and those of Mr. Pallone in focusing attention on this problem with the "Call Before You Dig" bill. Due to the increase in urbanization, an intricate web of cable, pipe, and optic fibers have been spun beneath our feet. While we generally need not fear where we step, the incident in New Jersey this past year shows us that we should be extremely concerned about where we dig. I agree with the goal of this legislation to establish a nationwide toll-free telephone number to protect all underground facilities from damage due to any excavation. This legislation will help us keep our underground facilities safe from inadvertent and damaging intrusions.

While this bill takes steps toward alleviating the problem of inadvertent cable cuts, I would like to point out that cable cuts are only a small part of a much larger problem. The integrity and reliability of these service facilities have been a concern of mine for years. Prompted by a rash of outages in 1991, the Telecommunications and Finance Subcommittee investigated the reliability of the communications network. One such outage crippled the New York airports leaving pilots without ground communication with the FAA center for 3 to 5 minutes and resulted in the worst interruption of aviation traffic in history. During the three hearings held on the matter in that year, some telephone company representatives insisted that the majority of these outages were the result of inadvertent cable cuts. At my behest, the General Accounting Office investigated the interruptions of telephone service.

The subsequent March 1993 GAO Report found that during 1990 and 1991, over 1,000 outages occurred, affecting over 69 million customers. Somewhat surprisingly, the GAO also found that only 15 percent—or 1 in 6—of all outages were due to cable cuts. In fact, the GAO found that 27 percent of all outages are attributable to hardware problems, 20 percent to software problems, and 15 percent to telephone company errors. Telephone company errors were almost as much as cable cuts.

The "Call Before You Dig" bill may protect the underground haven of our cable, pipe, telephone line and other facilities, but it doesn't even scratch the surface of the larger problem of telecommunications reliability. These phone outages have shown us that telecommunications are not only the economic lifeline of the country, but also its safety lifeline. Additionally, network reliability is crucial in light of the emergence of the information superhighway which will provide services such as up-to-the-second, life-saving medical information. So while I support the thrust of this bill, it only addresses one aspect of consumer service problems in these areas.

In conclusion, Mr. Chairman, I want to commend you and Mr. Pallone for bringing this important issue before the subcommittee. I share the goal of the chairman and the sponsor of the bill that we need to improve the safety and reliability of our infrastructure, and I stand ready to work with my colleagues to accomplish this goal.

Mr. SHARP. With that, let me recognize our two witnesses from the House of Representatives. We are very pleased first to hear from Mr. Pallone in whose district this terrible accident occurred earlier this year.

STATEMENT OF HON. FRANK PALLONE, JR., A REPRESENTATIVE IN CONGRESS FROM THE STATE OF NEW JERSEY

Mr. PALLONE. Thank you, Mr. Chairman, and I want to thank you for having this hearing and your attention to this problem, and I know it should be said that you have been very concerned about the safety issue with pipelines and particularly the need for a one-call system long before the accident that occurred at Durham Woods in Edison. This is not something that is new for you, certainly, and the rest of the committee also, I want to thank for their support and attention to the problem.

As you know, the Senators we expect to be here later and I just want to take a minute to recognize their help, as well as my colleague, Congressman Franks.

As you mentioned, the accident actually took place in my district which is in evidence, but Congressman Franks shares Edison with me, so even though I think he is maybe a few thousand feet away from the area where the residents were affected, we both share the municipalities.

I also wanted to recognize the efforts of State and local officials as well as the pipeline company, because people were really great in terms of pulling together and helping the victims of this accident. We were afraid, many people were afraid when they saw the occurrence on the evening news that a lot of people were going to be killed and we were going to find bodies and serious injuries. What actually happened was only one person lost her life who was fleeing from the accident, and it was really almost a miracle that happened, that so many people were able to get out, and a lot of it just happened because people helped each other.

Neighbors went next door and knocked on the doors and got people out who were disabled and it was fortunate we didn't have a lot more of a tragedy. I have to say that until a few months ago when this accident occurred, most of my constituents, including myself, had not given much thought to the issue of pipeline safety, and of course all that changed on March 23rd this year.

As you know, that was the night that the residents of Durham Woods Apartment Complex in Edison were awakened by this powerful explosion. Many were forced to flee their apartments on the heels of an intense fire that destroyed their homes.

I have to tell you, I was there, I believe the next day, and it was amazing to me, it just looked like a nuclear explosion. I mean, it is the type of devastation that occurred. It has been exactly 3 months since the accident and basically what we have been trying to do is to see what lessons have been learned in terms of help on the Federal or State level in terms of legislation, and that is really what I want to talk about this morning, the legislation.

One of the lessons that we learned, particularly from this explosion, is that like the majority of pipeline explosions, it was likely caused by third party damage, and the first bill that I introduced will address this issue by requiring that all States provide for a mandatory compulsory one-call notification system.

I know the gentleman from Idaho asked the question, why do we need Federal legislation? Well, let me say this. Most States, including our home State of New Jersey, have legislation that establishes some form of one-call system, but I would say the majority of the existing systems are inadequate.

First of all, they are voluntary, as is the case in New Jersey, and what this bill does, H.R. 4394, is to require the States to provide for one-call systems with mandatory participation for all excavators without exception or exemption. So it is the mandatory compulsory aspect that I think is different from most State's legislation.

It requires anyone who excavates to call before they dig and in order to allow excavators to obtain all the information about a site with one phone call, underground facilities used for gas, electricity, communications, water and sewage are required to participate. The bill provides for a strict system of penalties and allows for the issuance of citations, very similar to a traffic ticket for on the spot violations.

Excavators and facility operators who are not in compliance can be fined up to $25,000 but not less than $250 for each violation for every day the violation occurs, and the bill also provides for increased penalties in cases where the violation leads to death, serious injury or severe property damage. I believe that many of the accidents caused by third party damage will be prevented by this common sense measure.

Excavators will now be digging armed with the information that will allow them to excavate safely and those who refuse to comply endanger the lives and property of everyone around. The stiff fines will encourage compliance and punish recklessness, and I am confident that if passed, the bill will save lives and prevent the kind of destruction that we experienced in Edison.

As was mentioned by the chairman, on Tuesday I introduced a second bill, H.R. 4616, that addresses some specific problems that arose during the Edison accident that we also feel have Federal or nationwide implications. For example, one lesson that we learned in Edison is that pipeline inspections with smart pigs, as they are called, should be performed on a regular basis.

The Texas Eastern line in Edison had not been pigged since 1986 and it is possible that such an inspection could have detected the gouge that was found and likely the cause of the pipe break.

My bill requires a schedule for pipeline inspections, including inspections with smart pigs. The issue of remotely controlled and automatic valves also came to the forefront after the Edison accident. As you may recall, the pipe was equipped with a manual valve and it took several men almost 3 hours and 753 twists of the wheel to shut off the gas, which frankly surprised the pipeline company itself. They didn't realize it was going to take so long.

The bill requires that automatic or remotely controlled valves must be used wherever it is technically and economically feasible to do so.

There were also two siting issues raised. First, since Texas Eastern—since the line had been laid years before the Durham Woods Apartments were constructed, the question of development in areas where existing pipelines are sited arises and my bill requires FERC to prepare educational information for local governments to use for reference in making decisions on issuing permits for development in areas where existing pipelines are located.

Second, siting is an issue when new pipelines are routed through existing neighborhoods, and to address these concerns, the bill requires FERC to review their guidelines and procedures for siting in urban areas. I understand the agency has some reservations about this provision, Mr. Chairman, but I hope that the accident shows us that this is an issue that has to be looked at more closely, and I am also hopeful that we will be able to work with FERC to find a way to accomplish this that is not overly burdensome.

Finally, the bill takes steps to educate the public through a public education campaign to improve mapping, which Secretary of Transportation Pena was in New Jersey earlier this week and he stressed the need for a nationwide system of mapping because we don't know a lot of times where pipelines are located.

So it would improve mapping by instituting a program for data collection and it would take other lessons from the Edison accident and try to put those into legislation. I guess I should say that when the Secretary spoke at a summit that Senator Lautenberg put together earlier this week in New Jersey, he stressed a lot of things that are in this second bill, so hopefully we will be able to get his support and hear from him on some of the issues that are addressed in the second bill.

You know, Mr. Chairman, I am finished but I just wanted to mention, as you mentioned, that we have had other accidents since the Edison one. There is certainly not an isolated incident and I do think that this is the time to pass the one-call legislation and address the other issues and I really appreciate your attention to this.

Thank you.

Mr. SHARP. Thank you very much.

Mr. Franks. We are very pleased to have you now.

STATEMENT OF HON. BOB FRANKS, A REPRESENTATIVE IN CONGRESS FROM THE STATE OF NEW JERSEY

Mr. FRANKS. Mr. Chairman, thank you very much for your interest, as well as that of the committee. I would first like to commend my colleague, Mr. Pallone, as well as Senator Lautenberg and Senator Bradley for providing comprehensive leadership on this issue. I am here to underscore the fact that this is a bipartisan concern and a bipartisan interest. I commend Senator Lautenberg for his work, as well as the one-call bill under consideration today sponsored by Congressman Pallone.

I agree with Mr. Crapo's observation that this is essentially a very safe industry, but as you have noted, there have been accidents. In fact, there have been accidents subsequent to the Edison explosion.

In fact, between 1983 and 1992, 193 people have died due to natural gas pipeline explosions, and another 1,195 people were injured. These explosions caused over $31 million in property damage, and so I do think that despite the fact there is an enormous volume of gas transmitted, routinely on a very safe basis, we do have to understand the enormity of the power involved in one of these explosions, and I wish the one graphic that we might have been able to display today are pictures from Durham Woods because it literally looks like a nuclear explosion.

So I think any steps that we can take that encourage the States to have a more viable enforcement program, we can play a very helpful role in that regard.

As Congressman Pallone noted, New Jersey has had one of the better one-call systems and it was constructed without our help here in Washington and it has worked reasonably well, but I would note that this coming Monday, the New Jersey legislature is moving from a voluntary one-call system to a mandatory one-call system, and I suspect that other States will be moving in that direction.

As they pursue that direction, it seems to me to be important that we establish some national minimum standards that the one-call program should embody. To have a one-call system that is a patchwork quilt doesn't necessarily offer the greatest degree of safety to the residents of those States, nor the kind of continuity in programing that we could hope we could offer to the industry.

So I do think there is an important opportunity to construct an appropriate and reasonable Federal role in this process.

Too often, as Congressman Pallone noted, there are some deficiencies in one-call systems. They often exclude some types of excavators. The systems are fundamentally underutilized by contractors. They often have low fines and some very complex and protracted enforcement systems.

I think we need to address those issues and I think Congressman Pallone's bill is an excellent start in that process.

I did want to note that beyond the need for a one-call system, other issues clearly have been raised, and some legislation that I will be introducing with shared jurisdiction between this committee and Surface Transportation Subcommittee of Public Works on which I serve, we are going to be looking at some initiatives that, again, go beyond merely the one-call aspect to the potential solution here.

One is that I believe it is important to put real teeth in the enforcement effort, to make it a Federal crime to damage—to knowingly damage a natural gas pipeline and fail to report it. Again, the enormity of the damage that can be done and the risk posed to human life is such that I think this is an appropriate area to define as a criminal activity on a Federal basis.

Lastly, I think some questions on a longer term basis need to be addressed, questions which have been raised but never really been put to rest in the past, and that is to conduct an assessment of whatever risks to the public safety or the environment may be posed by natural gas pipelines.

Again, I think this has been an enormously safe industry with an excellent record, but it would be, I believe, helpful to conduct an assessment, a scientifically-based assessment that would help us draw some conclusions about the efficacy of this mode of transportation of natural gas and other hazardous materials.

Mr. Chairman, I again thank you for your time and the interest of the committee.

I want to again thank my colleague, Congressman Pallone, and hopefully the kind of tragedy that was visited upon the residents of Durham Woods won't have to be suffered by any other people in our country.

Mr. SHARP. Well, gentlemen, thank you very much. I don't have any questions. I will see if my colleagues do. We have already been communicating with both of you and both Senators and we will continue to do so as we seek to develop a consensus on legislation that we can move forward expeditiously, and many of these issues are not new, but we want to seize the opportunity to continue to make improvements in this area.

Gentleman from Illinois.

Mr. HASTERT. Just a couple of questions to clarify, Congressman Pallone. Was the Edison pipe that was ruptured or exploded, it was previously damaged; is that correct, and it was never reported? Or was it a new strike there?

Mr. PALLONE. The final report is not out yet, but the preliminary investigation showed that there was a gouge, so to speak, in the pipeline, and what we found is there has been a tremendous amount of activity in that area. In other words, first of all there was a lot of dumping on top of the pipeline.

Mr. HASTERT. Landfill?

Mr. PALLONE. Landfill and all kinds of material. Also, there was new construction in the area. There was a new development that was going up close to Durham Woods. So all indications are that the pipe was hit.

Mr. HASTERT. So your premise on this then, if there was good scrutiny prior to all this happening, that pipe wouldn't have been damaged.

Mr. PALLONE. I think that is accurate, but of course——

Mr. HASTERT. Indeed, in your premise then also is that you need to put in place those protections so that happens. Illinois has a pretty good program, the Julie program. It is mandatory. They have just had their five millionth call recently, but one of the things we found in Illinois is we put the penalties too high.

Other than a mandatory obligation to do it, you put the penalties too high and some small contractor is out there and he hits a pipe and, uh-oh, didn't call, better cover this hole up and get out of there, and it is also a detriment.

So you need to make sure that there is a balance, and if you get people tied into the mandatory call prior to anything being done, you are better off sometimes with it. High penalties is a detriment sometimes to making the calls even of the most serious nature.

The one-call minimum standard that you talked about, could States in your legislation exceed that?

Mr. PALLONE. I don't think there is anything that would prohibit them from doing it. This is a minimum standard in effect.

Mr. HASTERT. And then, one of the things that you talk about, barriers to new pipelines in populated areas, you know, I live in an area that is really a tremendous growth area, a lot of people moving in all the time, and of course as you say, that is where the most construction is because people who use natural gas, end users, are people who are building new houses and other houses.

So the alternative for that is propane, and propane moving over the streets in bottles and tanks and trucks, which is really less safe than something moving way under the ground, you know, and I think there are safety statistics that show that.

So we have got to be careful, too, in trying to prohibit safe delivery of this commodity that people demand and want and they are going get one way or another than having it shipped in heavily populated areas. You know, if it is in rural communities and it is not economically feasible to get that out there in a pipeline, they then haul it out in a propane tank. But to have thousands of propane tanks being moved through heavily populated communities is something that would be the alternative to this and I think something that we may not want.

Mr. PALLONE. I agree with you. I think that throughout the debate the last 3 months, we continue actually to bring up the fact that natural gas, or the transportation of natural gas, is one of the safest ways, so to speak, to transport energy, and that—you know, it is a difficult balance because as you say, you want new development, or at least most people do, exactly, and how do you provide the energy for new houses and new residential development.

Mr. HASTERT. And I just want to say, I look forward to working with you and Congressman Franks. This is a commendable goal and I think brings some balance to this whole issue and we would like to work with you on it further.

Thank you very much for your testimony.

Mr. PALLONE. Thank you.

Mr. SHARP. Mr. Crapo.

Mr. CRAPO. Thank you, Mr. Chairman.

I will just address my questions to both of you and you can decide who wants to kick in an answer. As I indicated, my main concerns are how we mesh the issue of Federal control with State sovereignty, and one of my first questions is, as I understand it, this would require that each State put into place a mandatory system, and establish the required penalties, or at least the minimum level of required penalties that each State must impose.

I also understand that there is a user fee involved at the Federal level for managing the Federal aspects of complying with this bill. How will the States pay for their participation in this program, or is that just left up to the States?

Mr. PALLONE. You know, I probably should check a minute. My assistant tells me it is left up to the States.

Mr. CRAPO. And the user fee, if you would just advise us. The user fee is the other bill.

Mr. PALLONE. You mean the second bill? There is no fee for the one-call system.

Mr. CRAPO. There is no fee involved for the one-call system, all right. Do we have any indication or any understanding of what kind of cost this will result on the State and local governments?

Mr. PALLONE. Well, again, I think that most of the actual cost is incurred by those who are using the system, and so I don't really see why, you know, the States should actually have any additional cost. It is pretty much left up to the discretion of the States as to how they want to raise the money.

Basically, they can charge a fee if they want to. That would be up to the States, or they can have the excavators pay for it in some way. There isn't anything that we put in the Federal legislation that specifies how—you know, how the funding for the program will occur.

Mr. CRAPO. OK. And let me ask just a couple questions about the penalties to make sure I understand how they work. As I read the bill it basically says there is a $25,000—or no more than $25,000.

Mr. PALLONE. Right. It is discretionary. Minimum I think I said was 250; then it goes to 25.

Mr. CRAPO. 250 to $25,000 per violation for each day the violation persists?

Mr. PALLONE. Yes.

Mr. CRAPO. Can you explain to me what a violation is?

Mr. PALLONE. The way I understand it, you know, a violation essentially, if you haven't called, if you haven't used the one-call system, or in the case of the utility, if they didn't get involved in actually doing the marking that they are supposed to do if someone calls and there is in fact a pipeline.

Mr. CRAPO. So in the case of the excavator, they must make the call. In the case of the utility, they must respond.

Mr. PALLONE. They must respond, right.

Mr. CRAPO. And is there a time line or is that left up to the State?

Mr. PALLONE. I think that is specified in there as something between 2 to 10 days after the—you know, after the——

Mr. CRAPO. The first call?

Mr. PALLONE. Right.

Mr. CRAPO. Now, if there is a call made and say 10 days elapses, is the excavator allowed to go ahead without liability or is that addressed in the act? In other words, what does an excavator do if they make the call and nothing happens?

Mr. PALLONE. And nothing happens. I don't think we address it actually, but that is a good point.

Mr. CRAPO. I guess—and then one other question I have is, if a call is made and a response is forthcoming that, no, there is no pipeline there, and then an excavation is made and there is a pipeline there, is the excavator then in violation for hitting the pipeline or is the utility in violation for not knowing that the pipeline is there, or I guess my question is, what if they didn't know and didn't determine? Is that going to then be a violation?

Mr. PALLONE. You are saying that neither the—in other words, the call was made, the utility company wasn't aware and I guess had no reason to be aware that there was a pipeline?

Mr. CRAPO. Yes. I don't know that would be a likely scenario, although in——

Mr. PALLONE. It is possible, because one of the problems that we have is that we don't have sufficient mapping, and a lot of times it is not clear that there is a pipe. So I mean, that is a possibility. But I would assume that is going to be just subject to your normal principles of negligence or whatever.

Mr. CRAPO. So the State basically would control the outcome of that circumstance?

Mr. PALLONE. One of the things that we are trying to do with this bill, and again, this is what the Secretary of Transportation stressed on Monday when we had this summit in New Jersey, is to try to improve the system, mapping system if you will, so we know where pipelines are, because there is definitely a problem.

I mean, I would say most utilities probably know when they put in a pipeline and where it is. But right now there is a definite problem in terms of local, State or even the Federal Government knowing and there may very well be instances where there is no knowledge. But I guess my answer to that would be that would just be left up to your normal tort law or whatever.

If someone has no knowledge and had no reason to have that obligation, then they might not be guilty—I mean they might not be liable.

Mr. CRAPO. And then just one last question. I come from an area that has a lot of rural country in it and I am not aware of any input from my own State on this issue. I am aware that in the State of Wyoming there have been concerns expressed about the financial impact of this type of a proposal on rural areas. Have you done any studies or are you aware of any analysis on the differential impact of this kind of a proposal on rural areas versus urban areas?

Mr. PALLONE. I am not aware, but I see that—I am getting a message here, wait a minute. OK, we actually have in the legislation that there might be some exceptions to time limit constraints on calling and making markings for remote areas, but you actually want to know if any studies have been done to see what the financial impact, whether it would be greater on rural areas versus impact? I don't have that, but perhaps we could find something or we could do some surveys.

Mr. CRAPO. All right. Thank you.

Mr. PALLONE. Thank you.

Mr. SHARP. Thank you. Gentlemen, we appreciate your concerns and your bringing them virtually immediately to our attention when the accident occurred and we applaud your work and we certainly look forward to working with you as we try to proceed on this issue.

Thank you very much for being here this morning.

We will now welcome our second panel of witnesses and what we will do is, when Senator Bradley or Senator Lautenberg arrive, we will let them come next in the order at the point that they come in, but we are very pleased to have with us again the Chair of the Federal Energy Regulatory Commission, the Honorable Elizabeth Ann Moler. We also have the Honorable Jim Hall, the Vice Chairman of the National Transportation Safety Board, Mr. George W. Tenley, Jr., Associate Administrator of the Office of Pipeline Safety, Mr. John F. Riordan, the chairman, president and CEO of MidCon

Corporation who is representing the American Gas Association and the International Gas Association of America.

I am sorry. We also have two more on this panel. Mr. Bob McMahan, the president of the Shell Pipeline representing the American Petroleum Institute and the Association of Oil Pipelines, and Mr. William P. Boswell, legal adviser for the One-Call Systems International.

Ladies and gentlemen, I think you are familiar with our processes. We will be happy to make your written statement and any written materials you brought with you a part of our printed record. We would appreciate your trying to hold your oral summary to 5 minutes so we can proceed expeditiously.

Chair Moler, we are very pleased to hear from you again.

STATEMENTS OF HON. ELIZABETH ANNE MOLER, CHAIR, FEDERAL ENERGY REGULATORY COMMISSION; JAMES E. HALL, ACTING CHAIRMAN, NATIONAL TRANSPORTATION SAFETY BOARD; GEORGE W. TENLEY, JR., ASSOCIATE ADMINISTRATOR FOR PIPELINE SAFETY, DEPARTMENT OF TRANSPORTATION; JOHN F. RIORDAN, ON BEHALF OF THE AMERICAN GAS ASSOCIATION AND THE INTERSTATE NATURAL GAS ASSOCIATION OF AMERICA; BOB McMAHAN, ON BEHALF OF THE AMERICAN PETROLEUM INSTITUTE AND THE ASSOCIATION OF OIL PIPELINES; AND WILLIAM P. BOSWELL, LEGAL ADVISER, ONE-CALL SYSTEMS INTERNATIONAL

Ms. MOLER. It is a pleasure to appear before the subcommittee once again today to discuss H.R. 4616, the Natural Gas Pipeline Safety Improvement Act of 1994 and H.R. 4394, the Comprehensive One-Call Notification Act of 1994, including your amendment in the nature of a substitute.

Natural gas plays a crucial role in the Nation's energy supply picture. It supplies 1/4 of our Nation's energy. Virtually everyone expects its importance and use to grow in response to our growing appetite for clean, environmentally desirable, reasonably priced fuel. I believe that the experience of last winter shows that natural gas is a reliable fuel that customers can count on even in severe conditions.

The pipeline transportation system is hidden from public eye. Statistics show clearly that the natural gas pipeline industry is very safe compared to other transportation modes. The public's acceptance of these critical transportation facilities stems from that safety record. Tragic accidents, like the ones in Edison New Jersey, Allentown, Pennsylvania and Wyoming jar us out of our complacency and remind us that pipeline safety is a critical issue for the industry and for the country.

As policy-makers we must take every reasonable step to insure that the Nation's natural gas transportation network is both safe and reliable.

I applaud the sponsors of the bills before the subcommittee for their very positive response to recent events. Clearly it is tempting to say "No more pipelines", or to adopt what has become known as the "BANANA" principle, that is Build Absolutely Nothing Anywhere Near Anybody. But the BANANA principle will not heat homes, cook meals, and fuel factories.

We must improve our safety record. The pending legislation meets that objective. I heartily endorse the One-Call program as proposed in the bill, as do all of the pipeline trade associations that we regulate. It is especially important that the program covers everyone who may do excavations and provide significant penalties as a deterrent against violators.

However, I do believe that one part of H.R. 4616 as mentioned in passing by Congressman Pallone earlier today is unnecessary. Section 7(a) requires the FERC to review its practices and guidelines for siting pipelines in urban areas and to determine whether changes are needed in the areas of route selection or the appropriate rights of way.

This section will not improve pipeline safety, an area over which the commission has no independent authority or expertise, but it will place a significant burden on the commission and provide a useless duplication of effort on the part of Federal regulators.

The primary responsibility for insuring pipeline safety resides with the Department of Transportation. Every certificate the FERC issues requires those who will construct pipeline facilities to comply with the DOT safety standards. The Office of Pipeline Safety has the experience and expertise to deal with safety issues. Its regulations are and must be the cornerstone of insuring pipeline safety.

We consistently apply those regulations and those criteria in our siting decisions. My prepared testimony outlines in some detail the very complex process that we go through, but I simply want to emphasize that we require pipeline owners and operators to abide by the DOT safety regulations, thus, I do not feel that the commission has any particular expertise to apply in this area and instead would prefer to defer to the expertise that the DOT possesses.

We have a very thorough-going, careful and time-consuming certificate process as it is. We consider safety to be a critical part of the pipeline gas transportation network and we understand that continued acceptability of these facilities in neighborhoods is important to our country.

I will be pleased to answer any questions the committee may have.

[The prepared statement of Ms. Moler follows:]

STATEMENT OF ELIZABETH ANNE MOLER, CHAIR, FEDERAL ENERGY REGULATORY COMMISSION

Mr. Chairman and members of the committee: I appreciate the opportunity to be here today to discuss H.R. 4616, the Natural Gas Pipeline Safety Improvement Act of 1994 and H.R. 4394, the Comprehensive One-Call Notification Act of 1994, and the amendment in the nature of a substitute.

Natural gas plays a crucial role in the Nation's energy supply picture. It supplies one-fourth of our energy. Virtually everyone expects its importance to grow in response to our growing appetite for clean,, environmentally desirable, reasonably priced fuel. And I believe that the experience of last winter shows that natural gas is a reliable fuel that customers can count on even in severe conditions.

The pipeline transportation system is hidden from the public eye. Statistics show clearly that the natural gas pipeline industry is very safe compared to other transportation modes. The public's acceptance of these critical facilities stems from that safety record. Tragic accidents like the recent ones in Edison, New Jersey, and Allentown, Pennsylvania, jar us out of our complacency and remind us that pipeline safety is a critical issue for the industry and for the country. As policymakers we must take every reasonable stop to ensure that the Nation's natural gas transportation network is both safe and reliable.

I applaud the sponsors of the bills before the subcommittee for their very positive response to recent events. Clearly it is tempting to say "no more pipelines" or to adopt the BANANA principle—Build Absolutely Nothing Anywhere Near Anybody. But the BANANA principle will not heat homes, cook meals, or fuel factories. Thus we must focus on efforts on improving our safety record. The pending legislation meets that objective. I heartily endorse the one-call program as proposed in the bill. It is especially important that the program covers everyone who may do excavations and significant penalties as a deterrent against violators.

However, I believe that one part of H.R. 4616, the Natural Gas Pipeline Safety Improvement Act, section 7(a), is unnecessary. Section 7(a) requires the Commission to review its practices and guidelines for siting pipelines in urban areas and to determine whether changes are needed in the areas of route selection or the appropriate width for rights-of-way.

This section will not improve pipeline safety, an area over which the Commission has no independent authority or expertise. But it will place a significant burden on the Commission and provide a useless duplication of effort on the part of the Federal regulators.

The primary responsibility for ensuring the safety of natural gas pipelines rests with the Office of Pipeline Safety within the Department of Transportation. That Office sets standards for the design, installation, inspection, emergency plans and procedures, testing, construction, extension, operation, replacement and maintenance of pipeline facilities. The Office of Pipeline Safety has the experience and expertise to deal with safety issues. Their regulations are and must be the cornerstone of ensuring pipeline safety.

By contrast, the Commission's siting decisions must balance many considerations. Safety in urban areas is very important to us, and so are the many other land use concerns we are charged by law to consider. However, there is simply no way that siting alone can be the principal guarantor of pipeline safety, because pipelines must sometimes traverse populated areas and new development will sometimes take place after a pipeline is built. Ultimately the key to assuring safe pipeline construction and operation is the DOT safety program.

More importantly, the Commission's existing certification process is both thorough and careful. It will not, in my judgment, benefit from the type of review contemplated in section 7(a). Simply put, there is no reason for us to substitute our judgment for DOT's safety expertise.

Let me describe our current procedures for analyzing whether a proposed project should be granted a certificate of public convenience and necessity under the Natural Gas Act. To begin with, any applicant for a pipeline certificate provides a great deal of information about the proposal—including detailed maps of the specific location proposed for the pipeline, environmental information about the location, and explanations of how it tried to use existing rights-of-way (for example, along electric power line routes).

The applicant must certify that it will design, install, inspect, test, construct, operate, replace and maintain the facilities in accord with Federal safety standards. That is, the applicant must agree to abide by all of the Department of Transportation's regulations. Any failure to meet this condition would be a violation of the certificate that the Commission would consider a very serious matter.

The applicant files an environmental report to support its proposal. This gives information about the environmental effects of the proposal and discusses alternatives. As part of this report, the applicant must identify the location of all residences and businesses within 50 feet of the right-of-way.

After an application is complete, notice is published in the Federal Register and all interested parties are invited to take part in the Commission's consideration of the application. In many cases, we also issue a specific environmental notice that asks the people who would be most affected by a proposal to give us input on the scope of the issues that should be considered.

The Commission independently analyzes environmental issues raised in the application and by intervenors and commenters. We analyze and verify the information submitted by the company, conduct our own studies and consult with others who have expertise on the resources affected by the proposal.

We publish our findings, including recommendations to mitigate impacts. As part of these findings, the commission may recommend alternative routings. This involves very detailed consideration of local terrain. Literally, it is a matter of poring over detailed maps of an area, conducting site visits—sometimes repeatedly, considering the possible alternatives and weighing the positive and negative effects of each one.

Again, the Commission requests public comment. And again it weighs whatever comments it receives. Only then does it decide to approve or reject the application, or to approve the application with modifications to mitigate any problems.

I want to stress two aspects of this process. First, we encourage public participation throughout the entire environmental review. This is especially relevant to issues of pipeline safety in populated areas. Public participation is one of the best ways to identify both potential problems and possible alternatives.

Second, the Commission's consideration of pipeline proposals involves many competing interests, especially in urban areas. These interests include not only safety, but also many matters specified by statutes like the National Environmental Policy Act, the Endangered Species Act, the National Historic Preservation Act and the Coastal Zone Management Act. We must also consider the concerns of other Federal agencies and State and local governments. All this is in addition to the basic public interest issues that direct our activities under the Natural Gas Act.

The nature of this activity places a high premium on experience and judgment. In my view, no set of rules could ever be detailed enough and sensitive enough to substitute for the expertise of a seasoned staff and the careful deliberation of an informed commission.We put a great deal of care into our pipeline siting decisions—and it is appropriate that we do so. Part of that care involves devising the best solutions we can in an immense array of specific circumstances. In effect, we conduct a full review of our practice in every single case. I believe that's the best, indeed the only, way to fulfill our many obligations.

The combination of full public participation, an experienced professional staff and the informed judgment of five Commissioners is the best safeguard that we will balance all the relevant factors fairly and accurately. I do not believe that the formal, general review of our approach proposed in section 7(a) would be appropriate or would help us do our job any better.

Thank you very much for this chance to appear before you today.

Mr. SHARP. Thank you very much, Chair Moler. I am very pleased now to welcome Mr. Jim Hall with the National Transportation Safety Board

STATEMENT OF JAMES E. HALL

Mr. HALL. Thank you, Mr. Chairman, good morning committee members. Accompanying me today is Mr. Charles Batten of our Pipeline Division, who is the investigator in charge for the Edison accident and would be available if the committee had any questions, specific questions on that accident.

The National Transportation Safety Board favors passage of H.R. 4394 and is pleased that several components for successful damage prevention programs are incorporated in the bill. These program elements require full participation by all organizations that operate buried facilities, whether public or private, and a true State-wide One-Call notification center with specific instructions specifying the conditions for its use.

It includes also monetary administrative penalties against violators, injunctive relief against habitual violators and imminently dangerous excavation operations, and criminal actions against persons whose actions violate the law and result in casualties or substantial property damage. In addition, it has requirements for public and excavator education about the damage prevention program.

The Safety Board believes that H.R. 4394 can be enhanced by establishing a shorter time frame for State implementation because most States already have the State-wide One-Call notification centers and excavation damage prevention laws. Also, States should be provided as much flexibility as possible relative to the organization structure and the day-to-day operations of One-Call notification centers, and the provisions applicable to one-call notification centers and those applicable to State governments should be clearly delineated.

With respect to broader legislation affecting pipeline safety, some of the issues that have been addressed in our recommendations and investigations. We believe that decisions concerning pipeline routing should include, the consideration of potential hazards to public safety should product release and ignite, the distance radiant heat may threaten buildings and occupants, as well as the actual blast damage consequences should be considered.

For liquid pipelines, the assessment should also be based on the properties of the most hazardous product transported, and the pipelines operating characteristics including shut-off capabilities and area topography. The frequency of natural gas and liquid pipeline internal inspections should be based on the risk posed by the pipeline to adjacent populations and environmentally-sensitive areas, and remote-operated valves should be required in densely populated and environmentally sensitive areas, and the installed valves must not reduce the ability to perform internal pipeline inspections.

Finally, public information on the hazards of nearby pipelines must be provided to the residents and prospective residents. For such information to be disseminated to all affected parties, there should be cooperative effort among communities, landlords and pipeline companies.

That concludes my testimony. Thank you, Mr. Chairman.

[Testimony resumes on p. 67.]

[The prepared statement of Mr. Hall follows:]

Testimony of

James E. Hall, Acting Chairman

National Transportation Safety Board

Good morning Mr. Chairman and Members of the Subcommittee. I appreciate the opportunity to appear on behalf of the National Transportation Safety Board to discuss the "Compehensive One Call Notification Act," H. R. 4394, aimed at preventing excavation damage to our Nation's infrastructure, especially buried pipelines, and other measures to strengthen pipeline safety.

Excavation/Third Party Damage

Excavation-caused damage to pipelines is the largest single cause of pipeline accidents today. Of the accidents reported to the Office of Pipeline Safety in the Research and Special Programs Administration (RSPA) in 1993, 27.1 percent of the 188 distribution system accidents, 16.5 percent of the 97 gas gathering and transmission accidents, and 22.3 of the 229 liquid pipeline accidents were caused by excavation errors. Of the accidents reported in 1992, 34 percent of the 103 gas distribution system accidents, 14.7 percent of the 75 gas gathering and transmission accidents, and 18.8 percent of the 223 liquid pipeline accidents were caused by third party damage. Property damage losses were more than $18 million in 1993 and $31 million in 1992. However, our review of reports to RSPA found that the dollar value of the losses caused by excavation damage and reported to the Office of Pipeline Safety are grossly underestimated. Although the narrative portion of accident forms describe damage at many accidents, the accident forms report zero dollar damage.

The effects of outside force damage accidents are most severe in the gas distribution systems. Those accidents account for most of the injuries and fatalities in gas distribution accidents and most of the environmental damages caused by product releases from liquid pipelines.

The Safety Board is consequently very concerned about this issue and welcomes this hearing. This morning we will discuss several accidents that highlight existing problems and demonstrate the need for new safeguards like those proposed in H.R. 4394 to reduce excavation, or so-called third party damage to pipelines.

On March 28, 1993, at Reston, Virginia, more than 400,000 gallons of diesel fuel was released into Sugarland Run Creek and the Potomac River, when a 36-inch-diameter petroleum products pipeline ruptured. Although no one was injured, there was substantial environmental damage. The Safety Board's investigation revealed that the pipe ruptured because of metal fatigue in a crack initiated by mechanical damage from heavy construction equipment at an undetermined time. Despite a thorough search of construction records for the seven years prior to the accident, we were unable to determine when the damage may have occurred.

Our investigation found that the pipeline company involved was an active participant in the area's one-call notification system. Moreover, our investigators found that the State of Virginia required excavators to give 48 hours notice in advance of any planned excavations.

Excavation damage also arose in an accident in June 1993, in Cliffwood Beach, New Jersey, where three residents died and three other individuals were injured when a house exploded. In this accident, a contractor working for the local gas distribution company informed the one-call notice center of the date when excavation was to begin. Excavation notification in New Jersey at the time of the accident was voluntary. Before beginning work, the contractor was provided information on the gas line's location, but not on the gas service line locations in the area. The contractor was installing new gas lines when his equipment struck and severed an unmarked, inactive steel gas service line. The contractor installed a temporary repair on the visible break but he did not check for other possible damage. The Safety Board found a second break in a 3/4-inch inactive service line at a fitting within a buried, paved-over foundation of a previously demolished house occurred. When gas pressure to the service line was restored, gas then migrated into an occupied house and exploded.

Another fatal accident occurred on July 22, 1993 in St. Paul, Minnesota, killing three people and injuring 12 others. A city public works employee severed a 1 1/4-inch plastic service line with a backhoe at a joint adjacent to a building foundation wall. Gas entered and ignited in the building. The Minnesota one-call notification center had been notified of the construction and gas company personnel had accurately marked the location of the pipeline before construction began. Minnesota State law required notice before excavating, and it applied to all excavators and all operators of buried facilities.

Three other accidents which have occurred this year further graphically demonstrate the ramifications of third-party pipeline damage.

On March 23, 1994, a 36-inch-diameter steel high pressure gas transmission pipeline in Edison, New Jersey, ruptured. The escaping gas ignited, sending flames hundreds of feet into the air. Heat radiated from the natural gas flames and ignited several apartment buildings. Eight apartment buildings were destroyed by fire, and more than 1,500 residents of the apartment complex were evacuated. Dozens of residents and some emergency response personnel were injured, treated at a local hospital and released.

On inspection of the pipeline, our investigators saw that the pipe had been damaged by at least two different types of mechanized excavation machines. The time of the damage is yet undetermined. Our investigators have learned that the transmission company had not been notified of any excavation activity in the area, even though the company was an active participant in New Jersey's voluntary one-call notification system.

In another accident investigated by the Safety Board, a 10-inch-diameter natural gas gathering pipeline near Green River, Wyoming, was ruptured on May 3, 1994, when excavation equipment operated by contractor employees struck and damaged the pipeline. The two employees operating the equipment were killed and two others injured. There was a one-call notification center operating in the area. However, the Wyoming damage prevention law did not require participation in the notification system and the pipeline operator did not participate in the notification system.

The most recent 1994 accident occurred June 9 in Allentown, Pennsylvania. In this accident, a 2-inch-diameter natural gas service line was pulled apart at a coupling less than four feet from an eight-story senior citizen apartment building. The coupling was separated on the side nearest the building by a distance of about 2.5 inches. Another coupling on the line exhibited signs of being pulled about a quarter inch, but it was not completely separated. There were numerous gouges and tears in the pipe's coating on a portion of a 59-foot section of pipe, and the pipe was permanently deformed and removed for further examination. One person died in this accident, one was critically injured, and almost 100 people were injured.

Adjacent to a portion of the pipeline, an old leaking fuel tank, its concrete pad and contaminated soil were removed by a contractor. The excavation was reportedly being backfilled. The pipeline was exposed in the excavation area, and testimony and pictures conflict as to whether the excavated line segment was being supported. The bucket for the track hoe was observed in the excavation, and the Safety Board is investigating to determine whether the pipe was pulled out of the coupling due to the pipeline being unsupported, being struck by the track hoe bucket, or as a result of the trench sidewall caving into the excavation.

Gas migrated from the coupling to the building's boiler room. A large vacant area existed under the boiler room floor to accumulate fresh air to the boilers. Grates were in the side outside walls and through the boiler room floor. There were vertical shafts from the boiler room to all floors in the building for electrical conduit and a trash chute. This allowed gas to go up through the building. Safety Board investigators found damage on all floors as a result of gas migrating through the building which ignited and exploded. Some of the most severe damage was on the fifth through seventh floors, and reinforced concrete on the bottom of the first floor was moved upward almost a foot in some locations by the explosion's force.

The type of accidents described above, unfortunately, are not new occurrences. Almost since the beginning of the Safety Board in 1967, we have investigated excavation-caused pipeline accidents, and based on these investigations, recommendations have been issued calling for remedial action.

The accident investigations in the late 1960s and early 1970s prompted the Safety Board to hold a symposium in 1972 on preventing excavation-caused damages. That symposium brought together representatives of organizations involved with excavation and pipeline operations to explore solutions to the problem and to improve public safety. As a result of the symposium and the subsequent recommendations issued by the Safety Board, the American Public Works Association today serves as the primary support for local, state, and national groups dedicated to preventing excavation damage to our Nation's pipeline infrastructure.

When we conducted our symposium, only two one-call excavation notification systems -- one in Portland, Oregon, and the other in Rochester, New York -- existed. All but three States currently have one-call notification systems, with most States providing statewide coverage through a single one-call center. Today, Utility Location and Coordination Councils composed of operators of underground facilities, excavators, and city, county, and state agencies now operate in most communities. These councils coordinate planned major construction projects to minimize pipeline damages. In 1972, only a few of these councils existed. A national council of ULCCs now exists to provide general guidance to many local and state organizations. There were no State laws on excavation damage prevention in 1972, but today laws exist in all but seven States.

Clearly progress has been made. The voluntary, cooperative actions of buried facility operators, state lawmakers, and excavators have been most responsible for reversing the increasing excavation-caused accident trend experienced in the 1970s. As a result of these safety advances, many lives have been saved, injuries prevented, and countless dollars in property damage avoided. It is relevant to note that these improvements came at a time when there was substantial growth in residential, commercial and highway construction, and extensive urban renewal projects in most large cities.

Once most of the States implemented one-call notification systems, activated underground location and coordination councils, and implemented damage prevention laws, the frequency of excavation-caused pipeline accidents markedly reduced. Unfortunately, that rapid reduction trend began to fade in the 1980s, and the downward trend has stagnated. The pipeline accident statistics since 1985 show that the average number of outside-force-caused pipeline accidents has remained relatively constant. But further decreases are needed as the discussed Safety Board investigations demonstrate.

Safety Board staff consequently initiated talks with the various accident prevention organizations to explore ways to enhance existing programs. In these discussions, staff learned that many of the original damage prevention objectives have yet to be achieved. For example, the full potential of one-call excavation notification systems have not been realized. This is because many buried facility operators, primarily government-owned water, traffic control, and sewer systems, do not participate in active on-call centers. Consequently, contractors and excavators who want to learn of infrastructure locations to minimize damage must not only call the one-call center, but also must individually call all non-participating operators. Furthermore, there are agencies and organizations performing excavations that do not bother with notifying anyone of their planned excavations. Safety Board staff learned that State agencies, railroads, and Federal county, and local government units often obtain statutory exemptions from notification requirements; yet, their actions pose the same threats to the infrastructure as do other excavators. Additionally, the penalty provisions in most state laws are not invoked because they are cumbersome and costly to administer. As such, nationally there have been few instances where a person has been penalized for violating any provision of state damage prevention laws.

The Safety Board consequently believes these programs must be strengthened and welcomes the introduction of H.R. 4394. Although our discussions are not finished, already we have identified several key components for successful damage prevention programs and we are pleased that many of these elements are incorporated in H.R. 4394. These program elements require:

- Full participation by all organizations that operate buried facilities, whether public or private;

- A true, statewide one-call notification center with specific instructions specifying the conditions for its use; and

- Monetary administrative penalties against violators, injunctive relief against habitual violators and imminently dangerous excavation operations, and criminal action against persons whose actions violate the law and result in casualties or substantial property damage.

The Safety Board favors passage of H. R. 4394 and offers for your consideration several ways the legislation could be further enhanced to afford additional protection for pipelines and the public.

Most States already have a statewide one-call notification center and have excavation damage prevention statutes that could be amended readily to bring state programs into compliance with the provisions in H.R. 4394. We therefore believe that too much time is afforded in H.R. 4394 for State implementation.

The legislation should clarify that the States are to establish excavation damage prevention programs, not one-call notification systems. As mentioned earlier, all existing one-call notification systems have been established through voluntary, cooperative relationships among operators of buried facilities. Their structures and operating conditions have been developed to address local conditions unique to the organizing agencies. The Safety Board believes that flexibility has been one of the more important factors in the successes achieved. We therefore encourage the Committee to retain as much flexibility as possible relative to the organization structure and the day-to-day operations of one-call notification centers.

To avoid confusion and duplication of effort, the provisions applicable to one-call notification center operations and those applicable to state governments should be clearly distinguished. This could be accomplished by using the term "excavation damage prevention program" when referring to actions and using the term "one-call notification center" when referring to the system used to receive and disseminate information on planned excavations.

There should also be a requirement for either the state or the one-call notification center to provide public and excavator education opportunities about the damage prevention program. Information on the state damage prevention statute and use of one-call notification system could help broaden its use. In this vein, training and licensing provisions for operators of mechanized excavation equipment should also be considered by the states.

Pipelines in Densely Populated Areas

The Safety Board has identified the need for additional safety standards for pipelines in densely populated areas. Our investigation into the Edison, New Jersey, accident discussed earlier is just one accident we are examining in densely populated areas. Another accident investigated in 1986 occurred in Mounds View, Minnesota.

84-088 O - 94 - 3

This accident involved the rupture of a pipeline and spread of gasoline and fire in parts of six blocks that resulted in two fatalities and one injury. In San Bernardino, California in 1989, gasoline was released from a pipeline resulting in two fatalities, 19 injuries, and damage to 17 homes. As a result of both the 1986 and 1989 investigations, recommendations calling for the installation of remote-operated valves in highly populated areas were issued. The recent explosion in Allentown, Pennsylvania is another example of the potential danger of pipeline gas and vapors when released in densely-occupied areas or near tall structures.

Liquid pipelines also traverse densely populated urban areas, and the close proximity of people to high-pressure liquid pipelines creates a need for similar or better safety standards. The need for routing, establishing rights-of-way widths, and providing educational information to assist local governments is even greater for liquid pipelines used to transport highly volatile liquids.

As an example of how far heavier-than-air gases might spread and the resultant impact is demonstrated by an accident in North Blenheim, New York. Propane gas escaped from a pipeline and flowed more than 1/2-mile from the rupture before it ignited, resulting in two deaths, seven injuries, and more than $3 million in damages. A liquefied petroleum gas release from a pipeline near Whitharral, Texas, engulfed a 3/4-mile by 1/4-mile wide area before it ignited, resulting in five deaths and two injuries. Obviously, in more densely populated areas hundreds of fatalities could result from similar accidents.

In past safety recommendations, the Safety Board addressed the need to educate localities and the public on land-use planning for properties in the vicinity of pipelines, especially for those pipelines used to transport highly volatile (HVL) liquids, materials such as liquified petroleum gases (LPGs) and anhydrous ammonia. In one specific accident report, the Safety Board concluded that a city plan did "not address the hazards posed by the proximity of mainline railroads and of high pressure pipelines." In this report, a recommendation to the City of San Bernardino in California asked for revisions in the existing land use plan to account for the location of high pressure lines. The National Association of Counties and the National League of Cities were requested to inform their members of the guidance contained in the National Research Council's special report entitled "Pipelines and Public Safety." An article on the topic did appear in a National Association of Counties publication and no response was received from the National League of Cities.

After the Edison accident, however, it is evident that questions on these issues remain and should be addressed. The Safety Board is aware that legislation is being drafted concerning these issues. Safety Board staff has seen several drafts of this proposal and believes several issues relating to lands adjacent to pipelines warrant discussion. These are:

1. Natural Gas Pipeline Routing -- When certifying new natural gas pipelines, the Federal Energy Regulatory Commission (FERC) requires pipelines to meet Department of Transportation (DOT) safety standards and, therefore, has a shared responsibility with DOT for safety. Decisions made by FERC about changes needed

for selecting pipeline routes and for determining the appropriate rights-of-way widths should include the potential hazards to public safety. The hazard assessments should be made not only of the blast damage consequences, but also on the distance radiant heat may threaten buildings and occupants should a pipe rupture and the escaping natural gas ignite. If danger (risk) zones are established, they should be based on specific criteria. These criteria are detailed in the next issue discussed.

2. Liquid Pipeline Routing -- DOT could establish rights-of-way widths and develop educational information to aid local jurisdictions. DOT Offices of Pipeline Safety and Hazardous Materials could follow review procedures similar to those used by FERC for natural gas pipelines to develop routing criteria and minimum rights-of-way widths for each of the liquid products transported by pipeline. Danger (risk) zones should be established based on the properties of the most hazardous product transported in the pipeline, the operating characteristics of the pipeline including the capability of the operator to promptly detect and stop the flow of product to the release point, the topography along the pipeline, and the existence of underground systems in which released material may enter and flow long distances from the release site.

3. Frequency of Internal Inspection -- The frequency of both natural gas and liquid pipeline internal inspections should be based on the risks posed by the pipeline to adjacent populations and environmentally sensitive areas. Factors that should be

considered in assessing those risks should include the incident, operating and maintenance histories of the pipeline segment, the properties of and hazards posed by the products being transported, and the maximum operating pressure.

4. Public Information -- People residing within the designated risk zones need periodic reminders informing them how to recognize hazardous conditions and what actions they should take for their safety. Since 1984, the Safety Board has encouraged the provision of information to prospective residents and tenants before they purchase or sign leases in the designated zone. This information would have to go to all tenants even though they may rent for less than a year.

5. Remote Operated Valves -- Since 1971, the Safety Board has repeatedly advocated the rapid shutdown of gas and liquid pipelines. For example, in the Mounds View, Minnesota accident discussed previously, remote operated valves were recommended based on the population at risk. After the San Bernardino accident, also mentioned earlier, this recommendation was reiterated and another recommendation issued to RSPA calling for the evaluation of check valves.

The shutoff valve issue is present in the Edison, New Jersey, accident as well. The force the employees needed to exert to operate the valve wheel became increasingly greater as the valve neared closure in this accident. This was due to the increasing pressure difference between the pressures upstream and downstream of the valve as the downstream gas pressure vented from the rupture which was just downstream of the valve. As the valve neared closure, the pressure differential dramatically increased and made it impossible for the crew to close the valve easily and quickly.

There are a number of alternatives to solve this problem, including back-up power valves, check valves and remotely operated valves. Check valves on one-way flow lines, in fact, are automatic closing valves. Used in combination with remote-operated valves, they often can provide the best means for rapidly isolating a failed section of pipeline and may be a potential cost-effective alternative. Neither check valves nor automatic valves should be excluded. In any case, the valves to be installed should not reduce the ability to perform internal pipeline inspections.

These are just a few of the improvements the Safety Board believes warrant attention in any pipeline safety protection initiative. Our accident investigations offer compelling testimony for further pipeline safety enhancements and we are willing to work with the panel in any way possible.

That concludes the Safety Board's prepared remarks and I would be pleased to answer any questions you may have.

Mr. SHARP. Thank you very much, Mr. Hall.

Mr. Tenley, we are pleased to hear from you now.

STATEMENT OF GEORGE W. TENLEY, JR.

Mr. TENLEY. Thank you, Mr. Chairman. I am honored to once again appear before you to discuss issues concerning one of the most important, yet least understood transportation systems in the Nation. I think we can encompass in the idea of a better and deeper understanding, knowledge, if you will, all that is yet to be done in insuring the integrity of this vital mode of energy transportation.

Clearly the concepts embraced in the legislation we will discuss today would add significant knowledge of this infrastructure, the risks to it and the risks it poses. In appearing before you today, I realize all too well that this is the last opportunity that I will have to work with you in framing a better, more capable pipeline safety program.

I truly regret this fact, for I as much as anybody, can appreciate what you have meant to the success that I believe this program is on the threshold of achieving. Your steady, reasonable, informed and often forgiving approach to the pipeline safety program has enabled us to progress even as we face an increasing set of challenges.

In both your personal leadership and in the quality and commitment of your staff, you have been instrumental in setting the course for where this program is going. All that follows will build on the foundation that you have set. I hope that as you go on to other challenges, you will rightly see the pipeline safety program as an important legacy of your career in Congress.

For the last several years, my colleagues and I in the Research and Special Programs Administration have worked with you and your staff and other authorizing committees in Congress to attack the problem of outside force damage, still the leading cause of pipeline failures, despite great strides that have been made over the last 15 years in the presence and quality of one-call damage prevention programs.

As you know, I have often spoken out about the fact that if we are to really take damage prevention, and its most effective instrument, one-call, to the next plateau of impact, we need to look at all underground facilities, not just pipelines. Clearly the bill under consideration seeks to do that and we are very pleased that the bill captures the essential components of what an effective One-Call system should look like.

There is a building consensus with all parties concerned about damage prevention that these elements, when working as part of a one-call center with state of the art technology, do make a difference.

As you well know, pipeline safety issues invoke the attention of a broad set of constituents, and consideration of One-Call enhancement broadens our constituent base even further. To be successful, we will need to form as broad a coalition as possible of various interests, both those who have a stake in protecting their systems and those who make their livelihood from activities that put underground facilities and the public at risk. The Department pledges to continue its commitment to work with the subcommittee and others to forge this coalition. For me, this is a real opportunity to help the pipeline safety program turn its focus away from crisis management to risk management.

The subcommittee is also considering specific pipeline safety legislation that would emphasize the concern of Congress over several issues that we in the Department are either currently considering or which it is likely our impending risk assessment model will identify as needing attention. As discussed more fully in my written statement, we fully support the provisions that seek to increase the knowledge of the public of the presence and rights of pipelines expanded grant authority to promote One-Call systems, new sanctions to address threats to the pipeline right-of-way, uniform computerized mapping that will support risk assessment planning at all levels of government.

As to other provisions that seemingly address current activities or the use of current authority, their presence in the bill sends the message to all of us involved in pipeline safety, government and industry, that this infrastructure and those who manage it will be held to a higher standard in the future. This is our new reality.

If we are to be successful in achieving what the public is increasingly demanding of us in ways that are effective and economically viable, we will have to work together on the pipeline safety agenda.

In bidding you farewell, Mr. Chairman, I reaffirm my commitment to that end.

Thank you.

[The prepared statement of Mr. Tenley follows:]

STATEMENT OF GEORGE W. TENLEY, JR., ASSOCIATE ADMINISTRATOR FOR PIPELINE SAFETY, RESEARCH AND SPECIAL PROGRAMS ADMINISTRATION, DEPARTMENT OF TRANSPORTATION

Good morning Mr. Chairman and members of the subcommittee.

I am George W. Tenley, Jr., Associate Administrator for Pipeline Safety, in the Research and Special Programs Administration (RSPA). Thank you for the opportunity to appear before you today to testify on behalf of Secretary Peika, Acting Administrator Ana Sol Gutierrez, and the Department of Transportation concerning key elements of a risk management regime to protect the Nation's pipeline infrastructure. In protecting the integrity of pipelines, we protect the health and welfare of the people and the environment in proximity to those pipelines.

The issues we will discuss today have been framed in legislation that has been proposed, or is under development, by your colleagues Congressmen Pallone and Franks, and in the Senate by Senators Bradley and Lautenberg. The bulk of my statement will focus on these legislative initiatives, but I believe it is important that I put these initiatives and the issues they confront in the context of the-pipeline safety program that will play a major role in carrying them out should they be adopted.

Mr. Chairman, as you know better perhaps than any Member of Congress, the Department's pipeline safety program has been in a constant state of flux for almost 10 years. With a growing list of mandates for action from Congress and the National Transportation Safety Board, most of which have followed accidents of national significance, and with resources incapable of keeping up with those mandates, the program has had to be one of crisis management rather than risk management. As such, we have established what I consider to be an admirable record for responding to accidents, while failing to have a program that adequately addresses prevention.

In the last year, we have witnessed accidents that raise fundamental issues concerning the state of the pipeline infrastructure and the need to better understand the risks posed to and by that infrastructure. However, I am pleased to report that, for the first time in the 12 years I have been involved with the pipeline safety program, I believe that the program will be equal to the challenges we face. While this is partly a function of the promise of more resources, I believe as important is the personal attention of Secretary Pena to those challenges. In my experience, no Secretary of Transportation has so informed him or herself of the issues involved in assuring the integrity and viability of one of the Nation's largest transportation infrastructures.

By committing the Department, in a time of severe budget constraints, to pursue a budget for pipeline safety in fiscal year 1995 that is double what it was in fiscal year 1993, the Secretary has signaled a new era for pipeline safety; an era in which the pipeline safety program can become a leader in risk management, and finally fulfill the essence of what its core purpose should be—a credible, technically solid, and independent assessor of the pipeline infrastructure and actions of the industry to protect and operate that infrastructure. I believe that the legislation currently under consideration in Congress can be an effective tool to help us realize that purpose if it avoids directing the means by which its outcomes are to be achieved.

Mr. Chairman, you have asked for our views on proposals contained in two bills: the "Comprehensive One-Call Notification Act of 1994" (an Amendment in the Nature of a Substitute to H.R. 4394; Discussion Draft dated June 16, 1994); and the "Natural Gas Pipeline Safety Improvement Act of 1994" (Discussion Draft dated June 15, 1994).

This bill addresses the single leading cause of pipeline failures, external force, or third party damage, typically caused by heavy construction equipment striking a pipeline. This damage may be acute, that is, it causes the pipeline to fail immediately and put construction crews immediately at risk; or, it may be chronic, that is, it causes damage to the pipeline that over time results in failure, thereby putting the nearby population at risk. It is this latter damage that is proving to be the more serious and difficult to control.

The bill is predicated on several concepts that we fully support, including: comprehensive state-wide one-call notification systems; a state pipeline safety grant program that encourages state adoption of effective one-call programs; the essential elements of an effective one-call system that features: mandatory participation by all underground facility operators, no exceptions for any type of class of excavators, effective sanctions, public education, and 24-hour coverage and provision for emergency excavation; development by the Department of a model one-call program; participation by the Department, through partnerships, in a nationwide public education program; and, the concept of a single dedicated telephone number for one-call notification, although we believe the single number could be applied statewide rather than nationwide in recognition of many existing state-wide numbers that have been advertised and in use for many years.

The bill also contains several provisions that, although we support in concept, we believe require further deliberation on the specific means by which they are to be achieved. These include:

—The bill's articulation of the elements of an effective one-call program. We believe the bill has accurately identified these key elements. However, we believe the bill's primary purpose of achieving a high standard of effectiveness nationwide could be achieved without *prescribing* the key elements in Federal law. While many state programs are not as far along as we would like, we have seen an increase recently in the willingness of state legislatures to improve one-call programs. This is due in large part to the growing awareness of the effectiveness of such programs, and the steady erosion of the objections of those who have previously failed to participate, such as municipalities and state, county, and local transportation bodies.

We have had discussions with representatives of successfully operating one-call systems who have indicated that a proper Federal legislative role would be provisions that describe the standards for effective one-call systems. We agree with this approach; and, as protection of the pipeline infrastructure is a principal goal of enhanced one-call, the legislation would support the Department's national campaign to achieve that end.

—The requirements for enforcement sanctions. It has been our experience that many state legislatures are unwilling to enact the large civil penalties required in the bill for excavators, and this requirement would preclude their participation in the program. It is more important that States have the flexibility to address violations of their one-call laws effectively, for example, by easily imposed administrative sanctions.

—Recognizing State-by-State variations in the federally-prescribed model. The bill seeks to avoid excessive Federal direction of state one-call programs by providing a process whereby the Department could review and approve state one-call programs that do not meet the bill's requirements. This process could add to the administrative burdens on the pipeline safety program, and place upon it responsibility to review aspects of underground systems for which we have no other authority or expertise (e.g., water and sewer systems, electrical and cable systems, and fiber optic systems). If the bill were more flexible by directly allowing State-by-State variations, and if the state reporting requirements were to include a periodic assessment of actions to bring States into conformity with the national model, then the Department, industry, and one-call centers could target their strategy for improving those state programs that are deficient.

—State reporting requirements. While supporting a provision that would require some state reporting, we are not prepared at this point to affirm the information requirements currently in the bill. We would like to work with the subcommittee and other concerned interests to identify the essential information that will enable sound decisionmaking and improved one-call systems.

—Grant funding for development of one-call systems. We support the concept of a grant program to encourage the development of effective one-call systems. However, we are concerned that the funding mechanism currently contemplated, namely use of $12 million in pipeline user fees collected but not yet appropriated, could have an unintended and adverse impact on the balance of the pipeline safety program. Because the pipeline safety user fees that fully fund the program must be collected after receipt of the annual appropriation, there is necessarily a gap between the receipt of that appropriation and the collection of fees for the year. Currently, the pipeline safety program can continue to operate because there are sufficient fees already collected. significant reduction of this funding cushion, particularly if the Department's requested increase in pipeline safety funding is approved, could result in delays in pipeline safety program activities because of shortfalls in funding.

Finally, the bill contains one provision that the Department believes requires further study. In order to assure that States adopt one-call programs that conform with the elements prescribed in the bill, States that do not adopt conforming programs within 3 years would have their highway trust fund allotment reduced by one-half of one percent for each year that the State does not adopt conforming programs. In an effort to assure that this provision has the desired result of stimulating state adoption of conforming one-call programs, the bill would authorize appropriations to the Secretary of an amount equal to the reduction that could be paid to the State for establishing a conforming program.

While the Department has successfully in the past used a similar approach to achieve state adoption of important safety initiatives under the Department's purview (e.g., 55-MPH limit; use of safety belts; child restraint systems; and 21-years of age alcohol usage limitation), this would reduce highway grant funds because of safety initiatives only minimally related to highway safety. The Department has had longstanding concerns with efforts to make use of the Highway Trust Fund for other than its intended purposes.

In sum, we look forward to continuing our work with the subcommittee and others in and out of Congress to craft legislation that will advance the breadth and quality of one-call programs.

With only minor technical adjustments that we are working with subcommittee staff to craft, we support the following provisions as particularly sound with respect to addressing the nature and impact of the Nation's pipeline systems:

—The bill correctly identifies the problem of the risk that is created by the coexistence of pipelines and people. We support the bill's assigning to the Department a role in providing educational information on pipelines to state and local land use planners is appropriate. The relationship of local land use decisions to pipeline risk has not been well understood, and we must improve this variable in the pipeline transportation risk equation.

—We support the provisions which would allow us to recoup direct costs of accident investigation from the pipeline company involved in an accident. This ensures the availability of sufficient resources for the oversight of post-accident corrective actions, with costs attributable directly to the responsible operator.

—We fully support expanding our authority to provide grants directly to the one call notification systems, and to allow the grants to be used to promote use of those systems.

—The provision which changes requirements for enforcement provisions in state one-call laws provides much needed flexibility to recognize differences in state practices.

—Development of uniform computerized mapping of pipeline facilities is a very important tool to enhance pipeline safety risk assessment and planning. We would work in partnership with our sister Federal agencies, States, industry and universities to assure a practical and secure system which can support decisionmaking on regulatory priorities, land use, environmental and emergency planning, and risk management generally.

—The Department fully supports the provisions for sanctions against those who would excavate and dump within a pipeline right-of-way, a risk we witnessed first hand in the Edison, New Jersey accident in March.

In addition to these provisions, the bill addresses issues that we do not believe need to be addressed in new Federal legislation, because we have adequate existing authority to deal with them, and in some instances are currently taking action to do so:

—In accordance with the Pipeline Safety Act of 1992 (PSA 92), we are currently developing a rulemaking on the periodic inspection of pipelines, under which we will determine the extent to which those inspections should be carried out with instrumented internal inspection devices ("smart pigs"). Under the mandate of PSA 92, we are currently planning to complete this rulemaking by October 1995. The bill seeks to amend PSA 92 to require completion within 1 year of enactment of the bill. Consequently, if the bill were enacted before October 1994, the completion of the rulemaking would be advanced by only a short period; if it were adopted after October 1994, the provision would be meaningless.

The provision would also amend the PSA 92 requirement by requiring the Department to prescribe the schedule for such inspections. Consistent with our ongoing rulemaking, we believe that we should set the risk-based criteria by which the operators schedule the inspections, and provide, if demonstrated to be necessary, minimum general time frames.

—In a similar fashion, the bill would require the Department to mandate, within 18 months and "wherever practicable", the installation of remotely controlled valves on natural gas transmission lines and hazardous liquid lines. With respect to the latter, we are currently developing rulemaking under PSA 92 that will address valves as an important emergency flow restricting device. That rulemaking is a key initiative under the Secretary's action plan, established to address issues arising out of the Colonial Pipeline company spill in Fairfax County, Virginia, in March 1993, for improving our environmental protection program.

With respect to the issues of improved valve operation and location for natural gas transmission pipelines, we currently have sufficient authority to address the issue. Our plan is to take what we learn from our ongoing comprehensive inspection of the interstate natural gas transmission lines in New Jersey, as well as the information we have gained from the Pipeline Safety Summit that convened earlier this week, and determine what actions we should take, including rulemaking, to address the enhancement of valve operation and spacing.

The bill also seeks to improve the public's knowledge of pipelines by requiring operators to disseminate certain prescribed information to people living within 1,000 yards of the pipeline. Operators are already required by our regulations to provide information to the people living adjacent to pipelines. Consequently, we see this provision as needlessly legislating an outcome that is already substantially required.

Mr. Chairman, the pipeline safety program is on the threshold of becoming what it has needed to be for many years. The Secretary has pledged his personal support to assure that the program reaches its full potential. I know I speak for our Acting Administrator Ana Sol Gutierrez when I add our pledge for the same outcome. To reach our full potential will take more than resources—it will take a new leadership approach that seeks out partnerships with States, local governments, public interest groups, and the industry.

The successful management of the risks inherent in pipeline transportation will require not only sound public policy decisions by all levels of government, but will require a wider and deeper understanding of pipelines—where they are, what they do, how they are maintained, and what risks they pose. If we are successful in nothing else in the "new office of Pipeline Safety", we must succeed in increasing public awareness of pipelines—and assuring that such awareness occurs before an accident.

Thank you.

Mr. SHARP. Thank you very much, Mr. Tenley.

Mr. Riordan, we will be happy to hear from you now.

STATEMENT OF JOHN F. RIORDAN

Mr. RIORDAN. Good morning, I am John Riordan, president and CEO of MidCon Corporation headquartered in Lombard, Ill. I am pleased to testify today on behalf of the International Natural Gas Association of America, INGAA and the American Gas Association, AGA, on pipeline safety.

I commend you, Mr. Chairman, for holding this hearing, as your retirement nears, for the conscientious leadership you have shown in attempting to develop our Nation's energy policy. I would like to submit my entire statement into the record and will briefly summarize the highlights.

INGAA and AGA represent over 300 interstate natural gas transmission and distribution companies operating in the United States, Canada and Mexico. Together our companies transport and deliver over 90 percent of the Nation's natural gas serving over 56 million customers. I also want to commend and thank Representative Frank Pallone of New Jersey for introducing H.R. 4394, a bill to establish mandatory State-wide One-Call systems.

For years the natural gas industry has been a strong advocate of legislation establishing mandatory, comprehensive One-Call systems in every State. I would like to submit a statement into the record signed by five natural gas associations supporting enactment of national One-Call legislation.

The industry would also like to express its appreciation for the excellent work of the Utility Location and Coordinating Council, the American Public Works Association and the Department of Transportation's Office of Pipeline Safety work over the years to develop One-Call systems throughout the Nation.

We urge you, Mr. Chairman, and your colleagues, to recognize these efforts and be careful not to undermine the progress that has been achieved. Specifically regarding H.R. 4394, we recommend that any comprehensive Federal One-Call legislation be carefully crafted to not create mandated Federal or State operator One-Call systems, but rather give the States guidance, direction and incentives to improve the existing systems.

Department of Transportation records clearly show that since 1970, outside force or third party damage has caused over 63 percent of the natural gas pipeline incidents and nearly half of the fatalities. Forty-six States have some form of One-Call program on the books. Of the 46 States, only 27 have mandatory programs and many of the programs have too many exemptions and no real means of assessing penalties for noncompliance.

INGAA and AGA support a minimum set of standards or guidelines that would allow States flexibility on establishing a comprehensive One-Call system. Among the elements we support are a comprehensive One-Call program in every State, a common toll free number, mandatory participation by all excavators, participation by all underground government and utility facilities, including pipelines, telephone, electric and cable wires, appropriate and forcible penalty provisions for noncompliance with the caveat that could either reduce or waive the penalties if incidents are immediately reported, and a comprehensive public awareness program.

It is encouraging to find our One-Call tenets embodied in H.R. 4394. Both INGAA and the AGA Board of Directors strongly support passage of this legislation including the principles I outlined.

With time running out in this Congress, it has been a pleasure to work with you, Mr. Chairman, and the other industry groups to build a consensus for swift passage of comprehensive legislation.

Regarding the issue of pipeline safety and the reauthorization of the Pipeline Safety Act, it is generally acknowledged that the overall record of the natural gas pipeline industry is excellent. The National Transportation Safety Board's annual analysis of transportation fatalities has consistently found that transportation by pipelines is the safest mode of energy transportation. Natural gas pipelines are in the business of providing safe and reliable transportation. Without safety and reliability, we will not have markets. And without markets, of course we won't have customers. While our industry has an excellent safety record, we are always working toward maintaining and improving the safety of our facilities.

The industry is presently reviewing and assessing its own safety standards and procedures. I was recently appointed to chair an INGAA board task force to review the interstate natural gas pipeline safety standards and procedures and report back to the INGAA board with recommendations for improvements.

The issues we are reviewing include risk assessment, pipeline integrity analysis, what technological advantages can be made to improve smart pigs, remote and automatic shut-off valves, and what enhanced measures we can take to improve public awareness.

The Pipeline Safety Act was reauthorized in the last Congress and it was thoroughly debated by this subcommittee. Under your leadership and the other committees of jurisdiction, the Act's current authorization expires in 1995. Next year's reauthorization is the appropriate time and forum to address any additional pipeline safety issues which may need to be addressed. We encourage the subcommittee to focus its precious time and resources towards efforts to pass comprehensive One-Call legislation and revisit the Pipeline Safety Act next Congress.

Finally, just a few thoughts about user fee funding for the office of pipeline safety. We support an increase in funding where there is a clear need, justification demonstrated and where the activity is not being duplicated by industry. However, all the activities of the Office of Pipeline Safety are paid for by user fees on transmission pipelines. These costs are ultimately borne by the consumers of natural gas.

If the new DOT proposed budget passes, user fees will increase from $43 to $98 per mile. This represents a 400 percent increase in user fees over 8 years. I believe it is time to revisit the 100 percent user fee funding issues and consider other methods of assisting to fund this important office.

In addition, some of the projects for which DOT requested funding appear to duplicate the efforts of some of the current programs. In closing, I want to again thank you, Mr. Chairman, and I want to promise you that you will have the assistance of the natural gas industry in passing sound One-Call legislation.

Thank you for having this hearing and I appreciate the opportunity to testify.

[The prepared statement of Mr. Riordan follows:]

TESTIMONY OF JOHN F. RIORDAN
ON BEHALF OF
THE INTERSTATE NATURAL GAS ASSOCIATION OF AMERICA AND THE AMERICAN GAS ASSOCIATION BEFORE THE HOUSE ENERGY SUBCOMMITTEE ON ENERGY AND POWER
JUNE 23, 1994

Good morning, I am John F. Riordan, Chairman, President and CEO of MidCon Corp., headquartered in Lombard, Illinois. I am pleased to testify today on behalf of the Interstate Natural Gas Association of America (INGAA) and the American Gas Association (A.G.A.) on two bills introduced by Congressman Pallone dealing with pipeline safety.

INGAA is a non-profit national trade association representing the interstate natural gas transmission companies operating in the United States and Mexico and the interprovincial pipelines in Canada. The American Gas Association is a national trade association comprising approximately 275 natural gas distribution and transmission companies throughout the United States, Canada and Mexico. Together our member companies transport and deliver 90 percent of the nation's natural gas from the wellhead to the burnertip through the nation's 280,000 miles of interstate transmission pipelines and 1.5 million miles of distribution pipelines. Our members serve over 56 million customers in the United States.

I want to commend and thank Congressman Pallone for introducing H. R. 4394, a bill to establish mandatory, statewide comprehensive one-call systems that may be state-operated or state-chartered. The natural gas industry has long been supportive of a mandatory, comprehensive one-call system in every state. I would like to submit a statement for the record that shows the strong support for one-call legislation by A.G.A., the Association of Gas Distributors (AGD), the American Public Gas Association (APGA), INGAA, and the United Distribution Companies (UDC).

The industry would also like to express its appreciation for the excellent work of the Utility Location and Coordinating Council of the American Public Works Association and the Department of Transportation's Office of Pipeline Safety over the past years to develop one-call systems throughout the nation. We urge Congress to recognize these efforts and to be careful not to undermine the progress that has been achieved.

We recommend that any federal comprehensive one-call legislation be carefully crafted not to create mandated federal or state operated one-call systems but rather give guidance, direction and incentive to the states to improve existing systems or create state-wide systems.

According to the Department of Transportation (DOT) data, outside force or third party damage has caused over 63 percent of the natural gas transmission and distribution incidents reported since 1970 and nearly one-half of the fatalities. Forty-six states and the District of Columbia have enacted legislation that basically supports a one-call notification process. Twenty-seven states make participation mandatory. However, many of these state programs contain numerous exemptions and some of these programs have weak penalty provisions or none at all while others do not have enough teeth to motivate excavators to use the appropriate one-call system.

The federal government should set minimum standards for the establishment of one-call systems and permit states flexibility in meeting these standards. Our industry believes certain principles should be incorporated in any one-call legislation:

- each state should be required to establish or sponsor statewide comprehensive one-call coverage;

- there should eventually be one common toll-free telephone number that any excavator is required to call prior to digging, perhaps a three-digit number such as 344 (DIG), a seven digit number such as 663-2255 (ONE-CALL), or a 1-800 number;

- participation in the program should be mandatory for all who conduct excavation activities, including, but not limited to, contractors, utilities, municipalities, state and federal government agencies and, of course, operators of underground facilities;

- participation should also be required for all government and utility operators of facilities located underground, including all pipelines, telephone lines, electric wires and cable;

- appropriate enforceable penalty provisions, which would include administrative as well as criminal penalties, should be imposed for those who fail to call before excavating, with flexibility to waive or reduce these penalties for those who report immediately that they have struck a buried facility; and

- an effective public safety awareness program should be implemented in each state or by the federal government.

We are pleased to see that these concepts are embodied in Congressman Pallone's legislation. Both the INGAA and the A.G.A. boards of directors strongly support passage of one-call legislation embodying the above principles. We pledge to work with Congressman Pallone and with you, Mr. Chairman, for passage this year of solid one-call legislation which the industry and the public need to protect the underground service infrastructure.

Addressing the issue of pipeline safety, it is generally acknowledged that the overall record of the natural gas pipeline industry is excellent. The NTSB's annual analysis of transportation fatalities has consistently found that pipelines provide the safest mode of transportation.

The basic business of pipelines is to provide safe and reliable transportation. Without safety and reliability, we won't have markets and we won't have customers. While our industry has an excellent safety record, we are always working towards improving the safety of our facilities. At the most recent INGAA board meeting, held in April of this year, I was appointed to head a board-level task force to review interstate natural gas pipeline safety standards and procedures and to report back to the INGAA board at its next meeting in July with recommendations for improvements. This task force is focusing its attention on risk assessment, pipeline integrity analysis and a review of Gas Research Institute and Pipeline Research Committee projects and priorities. We are looking at what technological advances can be made to improve the use of "smart pigs" and other nondestructive evaluation technologies, and the use of remotely controlled and automatic shutoff valves. We are also considering what enhanced measures can be taken to improve public awareness. We are developing an action plan with recommendations for the INGAA board. The A.G.A. board also addressed the issue of one-call and pipeline safety at their meeting on June 1 and resolved to support a comprehensive national damage prevention program similar to what I have previously outlined.

Our INGAA task force met with you, Mr. Chairman, as well as with Secretary Peña, his Chief of Staff Ann Bormolini, Acting RSPA Administrator Ana Sol Gutierrez, RSPA Administrator nominee Dr. Dave Sharma, and Associate Administrator for Pipeline Safety George Tenley. We will continue to work with DOT toward our mutual goal of improving pipeline safety.

Both INGAA and A.G.A. were represented at the DOT Pipeline Safety Summit earlier this week in Newark, New Jersey. Secretary Pena held this summit to "explore a range of pipeline safety public policy issues from risk assessment to the economic context in which pipelines operate in this country...with a goal to identify possible solutions to the numerous issues that confront (DOT), and to identify areas for future inquiry that will help (DOT) to direct our program in the future."

The Natural Gas Pipeline Safety Act was reauthorized in the last Congress and is due to be reauthorized next year. The issue of safety was thoroughly debated by this Subcommittee and the other Committees of jurisdiction at the time, and numerous rulemaking requirements were placed on DOT, including consideration of periodic inspections by smart pigs in high-density-population areas. Members and their staffs put forth their best efforts which resulted in a good solid piece of legislation. The DOT is still in the process of implementing certain aspects of this legislation. We believe that next year's reauthorization is the appropriate time and forum to address any additional pipeline safety issues which may need to be addressed. We encourage Congress to focus its efforts at this time on the single most important pipeline safety issue - national one-call legislation.

I would like to make a few specific comments on Congressman Pallone's natural gas pipeline safety bill. We can support the requirement in Section 4 for FERC to review its practices and guidelines for siting transmission facilities and to provide educational information to local jurisdictions. However, this provision does not fully address the concerns of existing pipelines. A primary concern is that communities are moving closer to the pipelines, which were originally cited in mainly rural areas. It is apparent that local governments have exercised little or no control over land use or zoning near these pipelines. The pipeline involved in the Edison, New Jersey accident had been there for over thirty years and the community over time encroached on its right-of-way. In fact, construction is now under way for an apartment complex adjacent to the pipeline very near the site of the accident. We need to encourage local governments to modify their land use planning and zoning requirements with respect to existing utility rights-of-way. Recommended guidelines from FERC and DOT will be helpful only if local communities act on them. For example, the *1988 National Research Council Special Report 219, Pipeline and Public Safety* contains many good recommendations which should be useful.

In addition, natural gas pipelines are often subject to restrictions regarding the maintenance of rights-of-way, including the frequency with which encroaching vegetation may be removed. We believe there is a significant safety benefit associated with maintaining an open and clearly visible right-of-way.

We strongly support legislation to prohibit dumping of materials on or beneath pipeline rights-of-way. We encourage both Congress and the states to enact strong penalties for anyone excavating for this purpose. Pipeline operators already send an annual mailing which includes a request to residents along the pipeline right-of-way to report any excavation or dumping. We are studying ways to further increase public awareness.

Pipeline operators have used smart pigs for many years as one tool to assist in assessing the integrity of their facilities. Smart pigs were created to detect potential corrosion of the pipe. Currently they cannot detect hairline cracks, especially longitudinally with the pipeline, or most types of third-party damage. Pipeline operators decide where and when to run smart pigs based on a multitude of factors that create an integrity analysis for each pipeline segment. Establishing a set schedule would be inappropriate, waste scarce resources and may, in fact, cause a pigging device not to be run in the pipeline segments with the greatest risk based on integrity analysis factors.

As stated above, smart pigs detect loss of metal and corrosion of the pipe. Requiring pigging of new or replacement pipe is inappropriate as there is no corrosion. New and replacement pipe is currently required to be hydrostatically tested to detect any material or weld defects.

With respect to the use of remotely controlled and automatic valves, industry has had operational problems with them in the past, primarily with relating to data feedback to a control center and variations in operational pressures that mistakenly close the valves and shut down a line. We are examining, along with our research agencies and DOT, whether any technological improvements can be made to assist us in our pursuit of a better, safer pipeline system and how soon those technologies can become available. For example, groundbreaking ceremonies were recently held at Battelle in Columbus, Ohio, a $10 million pipeline test facility to support technological improvements in smart pigs and to investigate other innovative methods of assessing pipeline integrity. It should be operational by this time next year.

While DOT regulations require pipeline operators to have public education programs, and all our members do so, we are developing stronger, more effective community awareness and public education programs that provide the flexibility to meet the needs of each pipeline system. In Section 10, we suggest that the area of notification be left entirely to the discretion of DOT. The points addressed in that section regarding the location of the pipeline, a request to report any excavation or dumping, and a central phone number for such reporting by residents near the pipeline as well as actions to take in an emergency are all provided now by pipeline operators in their annual mailings to owners, and by most pipelines to all residents near the right-of-way. However, I assure you we will be investigating how to deliver this message more effectively in the future.

We support adequate funding for the Office of Pipeline Safety where the need is clearly demonstrated and is not being duplicated by industry. However, all activities of that office are paid for by user fees on transmission pipelines that are ultimately borne by the consumers of the gas. If the new DOT proposed budget passes, user fees will increase from $43 to $98/ mile and we will have experienced over a 400 percent increase since user fees began eight years ago. I believe it is time to revisit the user fee 100% funding issue and consider other methods of helping to fund this important program. In addition, some of the research projects for which DOT requested funding appear to duplicate ongoing or earlier work. We all need to be assured that additional funding for OPS does not duplicate existing programs and optimizes these dollars to address the most important safety issues first.

In closing, I want to again promise you, Mr. Chairman, the assistance of the natural gas industry to pass solid one-call legislation. I also want to assure you that, while we have a sound safety record, we will continue to seek ways to improve the safety of our pipeline facilities.

Mr. SHARP. Thank you. We are happy to hear now from Mr. McMahan.

STATEMENT OF BOB McMAHAN

Mr. MCMAHAN. Mr. Chairman, thank you for inviting us here today. I am Bob McMahan, president of Shell Pipeline Corporation and chairman of the General Committee of the American Petroleum Institute, the Committee on Pipelines.

I am here today on behalf of the Association of Oil Pipelines, AOPL, and have with me Pat Corcoran, who is the associate director of AOPL. AOPL is an unincorporated trade association of 78 common carrier companies. They represent over 300 companies involved in all aspects of the oil and natural gas industry, including exploration, production through transportation, refining and marketing.

Together, these two organizations represent virtually all of the U.S. pipeline transporters of petroleum and petroleum products and a significant majority of the gatherers who gather oil and gas by pipeline.

Liquid petroleum pipelines are the safest mode of transportation in the United States. There are more than 220,000 miles of petroleum pipelines which together transport ¼ of all the intercity freight in the United States. A major cause of pipeline accidents has been third party excavator damage, an activity over which a pipeline operator has little control.

In order to minimize such third party accidents, present Department of Transportation regulations provide for routing of proposed pipelines around congested and residential and industrial areas whenever possible. However, many pipelines were constructed years ago in rural areas which have since become populated.

Despite regular aerial, and in some cases surface, inspection of pipeline rights of way, above ground signs marking pipeline routes, active public education programs regarding the hazards of digging around pipelines and existing One-Call systems, third party damage continues to be a major problem.

One-Call systems have become a proven method of alerting the excavator to the existence and location of underground facilities and have been used effectively. It is and has been our position that One-Call systems should be encouraged, supported. Broad participation by pipeline companies, underground utilities, public works, highway departments and excavators alike should be mandatory.

With respect to H.R. 4394 that is before the committee today, we understand that an amendment to this bill, Mr. Chairman, has been offered and our comments are directed to that amendment.

Let me first say that the industry strongly supports the One-Call concepts and system. With regard to the specific questions that the committee had asked us to respond to, we do support broad-based participation in One-Call systems by both owners of underground facilities, including hazardous liquid pipelines regulated by the Department of Transportation, and excavators. We are especially pleased that under this legislation, government entities and public utilities, as well as private underground facility owners, are required to participate in the One-Call systems.

As to the benefits of a nationwide toll-free telephone number, we understand that what is envisioned is a single number in use nationwide that would access local One-Call systems, much like dialing the 411 access for the local directory assistance. We strongly support such a single number to facilitate communication and public education, including with that public service announcements and notification to people along the right-of-way, thereby improving the use and effectiveness of One-Call systems.

As previously indicated, the industry has existing public awareness programs. We support an aggressive education program to further the use of One-Call systems. Initiation of a government sponsored public education program we think is also a very positive approach.

With respect to the proposed penalties in enforcement, legislative language should consider the balance between encouraging participation and focusing on enforcement on repeat offenders. We don't want to make the penalties so large that they impede the bill's goals of improving communication between excavators and facility operators.

Presently some form of One-Call notification system exists in 48 States. Some of these systems are mandatory, while others are voluntary in nature. Most One-Call systems are formed by nonprofit organizations consisting mainly of representatives of the underground facility owners. These systems are effective, cost efficient, and as it is envisioned in the bill, should continue to operate.

We strongly support the ability of States to petition the Secretary of Transportation for a waiver with respect to One-Call systems that meet the minimum requirements of the legislation. Shell pipeline currently participates in 17 One-Call systems in 16 States and we actually have people participating on committees in all those One-Call systems and have 5 of our employees on the boards of directors of those 17 systems.

State agencies, we feel, should not operate the One-Call systems. They should only be responsible for providing guidelines and oversight and enforcement.

With respect to State grants, we understand that funds contemplated for grants to States to establish One-Call systems are currently funds in escrow and further appropriations should not occur until the next session of Congress when pipeline safety reauthorization is again considered.

Let me just comment briefly on the legislation, H.R. 4616 which Congressman Pallone referred to this morning. We have seen a draft version of that and just let me make a couple of brief comments relative to that. Basically though, because of the timing, we have not had an opportunity to thoroughly look at all these proposals and will make additional observations when we testify, but we—and we request an opportunity to provide further written comments relative to the pipeline safety legislation.

Our initial general appraisal of some of these issues are that some of them may be unnecessary and onerous and punitive in some cases on responsible pipeline operators. A particularly troublesome concept is the one that requires a company, regardless of who is at fault, to reimburse the Department of Transportation for costs the DOT incurs with respect to an incident at one of the company's facilities. Funds are already provided through user fees paid by the pipeline companies to underwrite the Office of Pipeline Safety.

On the subject of smart pigs, the Pipeline Safety Act of 1992 provided for periodic inspections of liquid pipelines in environmentally sensitive and high density population areas by amending the hazardous liquid Pipeline Safety Act. We understand that a further amendment is proposed to section 203(k)(2) which would provide that the Secretary establish a schedule for such inspections when the Secretary has issued regulations prescribing the use of smart pigs. It is our opinion that the Secretary already has the wherewithal to implement the provisions relative to smart pigs in the new safety bill.

On the subject of public awareness, regarding public awareness, it has always been a high priority with the pipeline industry. We understand legislation is being offered that would require operators to contact local residents within a specific distance of the pipeline. We agree that would be appropriate for the Secretary of Transportation to determine the distance within which notification should be required, however the creation of some arbitrary distance, such as a half a mile on each side of the pipeline, would be difficult, if not impossible.

Further, it is—we feel that the determination of that distance ought to be based on actual and realistic risk assessments.

Finally, we are concerned that the effectiveness of any such mandate that requires annual contact for every individual within a contemplated area would be almost impossible, particularly when that area can be as much as a mile wide.

Regarding the subject of the electronic gathering of information, this provision in the bill permits the DOT to develop an electronic database of information on the nature, extent and geographic location of pipeline facilities. Among other purposes, the database would be intended to assist States with land use planning, environmental and pipeline regulatory oversight, as well as risk management and safety planning with respect to such facilities.

We note that it is the—that the Office of Pipeline Safety has the primary responsibility for pipeline safety and regulatory environment issues, not State agencies.

Further, we question the incorporation of land use planning on a national basis because the complexity in the rapidly changing nature of this information, not to mention that most land use planning is performed more at the local township and county level, not by State agencies.

Comprehensive electronic databases are extremely expensive to develop and require continue updating for accuracy. Any company, State or other entity that benefits for a variety of purposes from such a program should be required to equitably bear the costs, rather than imposing another financial burden solely on the pipeline operators.

Regarding the one-call provisions, we would suggest that those provisions be handled under the bill, H.R. 4394.

Regarding additional funding to the Office of Pipeline Safety, basically the industry does not object to additional funding of the Office of Pipeline Safety where a clear need is demonstrated. But stated further, appropriations should only be considered in the next session of Congress when the pipeline reauthorization is up, when their efficacy and the amount can be scrutinized.

Mr. Chairman, that concludes my testimony. It is a pleasure to an appear before the committee today and I am here to answer any questions you may have.

Mr. SHARP. Thank you very much. We now hear from Mr. Boswell.

STATEMENT OF WILLIAM P. BOSWELL

Mr. BOSWELL. Mr. Chairman and members. Good morning. I am William P. Boswell, vice president and general counsel of the People's Natural Gas Company and Legal Adviser of OCSI. With me is William G. Kiger, executive director of the Pennsylvania One-Call system. We represent OCSI which stands for One-Call Systems International, which for 20 years has been the umbrella organization for one-call information, training, and education efforts throughout North America. OCSI is a part of the American Public Works Association which we also represent. We appreciate the committee invitation to make some general remarks at this time.

The one-call concept itself has been around for 30 years, although most of our efforts have occurred since the mid-1970's. Currently there are one-call systems in virtually every State except Hawaii. Participation is mandatory in over half of the States and the systems take some 25 to 30 million calls each year and turn them into 200 to 300 million messages to operators of underground facilities regarding proposed excavations. If we leave you with no other message, it is that one-call is working and it is working well.

Federal legislative initiatives and action have been limited to this point, although several Federal officials have been intimately involved by providing support, encouragement and some regulatory direction. The greatest service that the Federal Government could provide at this stage is to identify and encourage the States to implement the most useful aspects thus far developed.

Many States have programs in place which, with minimal modification, would have little trouble complying with the enumerated requirements. The exact parameters of a State's program should reflect the particular concerns that each State considers to be most important. Nevertheless, there ought to be relatively uniform approaches to certain matters, one which is reflective of common expectations on the part of those who operate facilities or excavate in various States.

We identify four key principles for the effective operation of a one-call system. They are mandatory membership, a duty to call, continuing education, and a graded enforcement system. Specifically, this includes the following: All owners and operators of underground facilities serving the public must be members of a one-call system. This applies as well to the State departments of transportation and municipalities which often have been exempted.

All excavators should be required to call before excavating. There should be one toll-free number for excavators to use to notify members of one-call systems in the geographic area of construction, which is preferably the State. The one-call notification service should be accessible 24-hours-a-day, 7 days a week.

Any one-call system must provide for a method by which members are timely notified. The one-call system should inform the caller of the identity of those being notified, give the caller a permanent file or serial number as a record of the call, voice record the call and retain the record for the State tort statute of limitations period.

Notifications should be made within a reasonable period of time to permit information to be required prior to the planned start of excavation. However, the notification should not be made so far in advance of excavation that the information provided regarding the underground facility becomes outdated. Emergency work which would permit a variance to this rule should be defined.

The one-call system should have or participate in documented continuing education program to inform the excavating public of the need to call before digging. The costs of operating the system should be recovered from those who make use of it. This may include the one-call system members and excavators.

There should be administrative, civil and criminal penalties for failure to comply with the particular State's program.

Now, penalties for not calling are necessary, but only as a last resort and then only in relation to the offense. Administrative sanctions generally are more effective than civil or criminal penalties because of the ease of administration. Getting people to call is the most important consideration. Getting people who haven't called to call if they hit a facility is the second most important consideration.

Specifically, penalties should never be so large nor so inflexibly applied as to discourage people from calling. An excavator who hits a line should not be so terrified to report the hit that he covers it up and hopes for the best. We don't need to prove the law of unintended consequences by having penalties drive behavior in negative ways.

Federal law should require that the States have a market basket of penalty provisions suitable to the offense. The penalties should move from administrative sanctions to civil fines and criminal sanctions. Of these, administrative penalties are the most effective because they are swift, easy to administer and they get the point across. Moreover, civil and criminal enforcement is handled through State attorney general offices and through local district attorneys. They have many other significant responsibilities and not normally will take a case unless there has been loss of life, significant personal injury or significant property damage.

We believe that the Federal Government can play an important role in this process. First, it can recognize that much of what happens has and will continue to happen at the State level, thus it should encourage the States to find ways of improving the damage prevention systems already in place.

It likewise should encourage the States to have systems which reach broad uniform goals. These previously have been stated by me, but to repeat, they are mandatory membership, a duty to call, continuing education and graded enforcement. Through the designated Federal agency, the Federal Government ought to work with interested parties to produce a model one-call damage prevention program. It should assist the States in one-call systems in developing pilot projects to improve their delivery of services, and the

Federal Government should actively pursue public education as the best means of getting people to call and respond.

We would be happy to answer any questions you might have, Mr. Chairman.

I would make two other points. Generally with respect to the comment regarding pipelines not knowing where their pipes are, pipelines know where their pipes are. The problem in our experience, and I have been doing this now for 16 years at the State level, is generally with water and sewer authorities, which are State agencies or local municipal agencies. They have a hard time knowing where their facilities are. Interstate pipelines which transmit natural gas know where their pipelines are. That is not the problem.

Another question that was asked by Mr. Crapo: Generally, a nonresponse means that the utility notified has no facility there. Some States do require a response, both positive and negative. If you don't have something there, you still need to call. If you do have something there, you have to come out and locate the facility itself, but normally if a call is made and there is no response within a particular period of time, the contractor can begin excavation.

Thank you, sir.

[Testimony resumes on p. 106.]

[The prepared statement and attachments of Messrs. Boswell and Kiger follow:]

TESTIMONY
OF
WILLIAM P. BOSWELL & WILLIAM G. KIGER
BEFORE THE ENERGY & POWER SUBCOMMITTEE OF THE COMMITTEE ON ENERGY & COMMERCE, HOUSE OF REPRESENTATIVES
UNITED STATES CONGRESS
REGARDING COMMENTS ON LEGISLATION HR 4394
AS INTRODUCED ON MAY 10, 1994
WASHINGTON, DC
JUNE 23, 1994

Good morning. I am William P. Boswell, and I appreciate the opportunity to present testimony prepared by William G. Kiger and myself regarding HR 4394, which, among other things, would mandate the States to implement one-call systems as a part of their damage prevention programs for underground facilities.

By way of introduction, Mr. Kiger is executive director of the Pennsylvania One Call System (POCS), a Pennsylvania not-for-profit corporation, and chair of the legislative committee of One Call Systems International (OCSI). OCSI is a part of the American Public Works Association, and is the umbrella organization for educational and other purposes for the one-call centers throughout North America and those other areas of the world where they exist. He is a past chair of OCSI, and helped found the organization in 1975.

I have been legal advisor to OCSI since 1980, and I am legal advisor emeritus and past chair of POCS. I am vice president and general counsel of The Peoples Natural Gas Company, Pittsburgh, Pennsylvania. Of some relevance on this subject, I also am an elected public official, serving as president of the borough council of Osborne, Pennsylvania. I have written, spoken, narrated videos, and otherwise been a key, volunteer player in the one-call effort throughout the country for more than 16 years.

HR 4394 is an excellent proposal from which to work toward a constructive federal role in a process that primarily has been addressed at the State level for the past 30 years. We emphasize that one-call damage prevention programs are widespread and that they work exceedingly well. All but two States have some form of one-call system (this will drop to one State as of 1/1/95), and all but four States (three as of 1/1/95) have damage prevention statutes. Collectively, OCSI members received more than 20 million calls to our centers in 1993, which generated some 300 million message notifications to facility owners and operators. These messages, called-in and received, dramatically reduced the risk and incidence of damage to persons and property.

One-call systems are constantly evolving and improving, but one thing is certain: the excavator must make the call, and the facility owner or operator must do something with the call it receives. Each of these actions is more likely to occur with proper education of the parties concerned.

With some relatively modest changes, HR 4349 could help move damage prevention efforts toward greater predictability from State to State, while continuing to give each State the flexibility it needs to experiment and to deal with local problems in the most effective manner. We urge that any action by Congress recognize both the value of what already exists, and be in support of State one-call efforts, which have been in place for more than a quarter of a century.

As a general proposition, we favor uniformity in certain specific areas.

First, all owners and operators of underground facilities serving the public should be required to be members of one-call systems, and to participate in damage prevention programs.

Second, those who excavate should be required to give advance, timely notice of their intention to do so, and they should be able to give such notice as easily as possible.

Third, there should be continuing education programs, the object of which would be to advise persons of the risks inherent in excavation, and of their responsibilities to minimize those risks by complying with the notice and other requirements of State law.

Fourth, where voluntary compliance does not work, there ought to be several means of enforcing notice and participation mandates in a timely and constructive fashion; these means should include administrative sanctions, injunctive relief, and civil and criminal penalties, in that order.

I. Participation. Traditionally, the owners and operators of underground facilities, and the large contractor organizations, have been key supporters of and participants in one- call and damage prevention programs. Much of this interest is related to the fact that it is their facilities, their equipment, and their employees who are at the greatest risk of damage and injury. Conversely, there are groups which traditionally have resisted participation; usually these have been the State departments of transportation and organizations representing local municipalities. It is ironic that these two groups are among the larger excavators and are themselves facility owners, with many underground facilities located within their public rights-of-way.

Often, the reasons cited by State DOTs for their reluctance to participate are these: first, they claim they never cause any damage; second, they claim it would be too inconvenient to their work schedules if they had to call ahead of time; or third, they claim they already know where all of the underground facilities in their jurisdiction are located.

Unfortunately, empirical evidence shows that these assertions are not correct. Traditionally, State DOTs have been exempted by State statute, in many cases as a result of the threat of a gubernatorial veto of legislation establishing a one-call program if they are included. As a half-way measure, some State DOTs have agreed to a requirement that outside contractors working for them, but not their own (State) employees, be covered by the law and thus required to call. In our opinion, there is no logical reason for a State DOT to be excluded; its own facilities, the facilities located along its rights-of-way, its employees, and the public are worthy of protection.

Municipal organizations object to participating for one of two reasons: first, that the cost of membership is something they do not wish to bear; and second, that if they become members and thus receive notice of intent to excavate, they will have to have someone respond to the call and locate their facilities.

As to the former objection, the cost of membership is nominal to most municipalities, normally consisting of the installation of a PC or a fax machine (which most of them have already), or something as simple as answering a telephone during normal business hours, and some one-call systems subsidize membership for small municipalities. Those municipalities with no underground facilities don't need to be members.

As to the latter, it is true that if you receive a call you have to do something about it. But the call, for example, can be tied into the excavation permit process, which most municipalities have, and this can be an excellent way of increasing permit revenue to a degree which will more than offset the cost of participation.

For instance, when the City of Pittsburgh, Pennsylvania joined the PA One Call System (POCS) its street opening permit revenue was about $7,000 annually. Currently, as a result of tying permit application and one call notification together, the city generates in excess of $350,000 annually in revenue at a cost of about $13,000 for POCS membership - the cost based on nothing more than the number of calls it receives from excavators.

There is no compelling argument against participation in a one call system, and there are, at present, creative, workable solutions which can make participation not only painless but even profitable to governmental facility owners and operators.

II. Notification. Those who intend to excavate, as that term is broadly understood, should give advance notice of their intent to do so, except in an emergency where notice is not practicable. Even in an emergency, however, notice should be given as soon as possible.

The notice should be given sufficiently in advance of excavation so as to permit affected parties to come to the job site to locate their facilities prior to excavation. Normally, this means that the notice should be given two or three business days in advance. The notification process ought to be easy to use for a project designer or excavator, and this generally should be through the use of a toll-free or "800" number directly to the one-call system serving the area where excavation is to occur. (Establishing a national one-call telephone number is a useful goal; however, and it should be pursued in stages. Certainly there ought be a national information number to permit excavators to ascertain the correct "800" number for the state in which they intend to excavate, and each state should have such a number, which should be **reachable** from any other state.)

Further, the excavator, when calling, must give specific information as to the job site and the nature of the work to be done. Finally, the excavator then must begin work, following the location of any underground facilities, in a timely fashion so that the any location marking does not become erased, stale or outdated due to the passage of time. The time-frame in the bill is appropriate.

III. Education. No one-call system will work unless it is used. Moreover, it misses the mark to consider one-call as anything other than an important part of an overall damage prevention program. The keys are to get excavators and facility owners/operators to become aware of their obligations, and to get them to see that it is in their own interest, as well as that of their employees, and the public, that they participate.

One-call systems and those who support them have comprehensive public education programs, which have been underway for years. Meetings with contractor groups, public service announcements, advertising media, presentations at trade shows, public proclamations, and virtually anything else which can be envisioned have been used. Some systems have regional representatives who go from place to place meeting with groups of excavators and facility owners/operators, and making presentations throughout the year.

OCSI has had an annual symposium on the subject since 1975, and this year's symposium, held in April in Minneapolis, attracted some 650 attendees over a five day period. Participants included one-call system operators, facility owners and operators, contractors, insurers, and federal, state and local officials.

The Federal agency given jurisdiction in this area should have a role in the public education process. We have found over the years that if people are educated they generally will participate, because intuitively the whole process makes sense. It makes sense from the standpoint of safety; it makes sense from the standpoint of saving time in the long run (quite apart from safety, if you don't call, begin construction, and find an underground facility, everything comes to a halt until you figure out what to do about it); and it makes sense from an economic standpoint (construction delays cost money, as does litigation).

Those who do not call find insurers lose patience and cancel's one's insurance. Without insurance, contractors can't bid on public contracts, and many private parties won't hire them either. It's a quick, painful way to go out of business. Not surprisingly, the insurance industry is a supporter of one-call damage prevention programs. So are responsible contractor organizations, as it's a good way of getting rid of competition from non-responsible excavators, as well as the bad press from the damage caused by those who don't call.

IV. Enforcement Mechanisms. Education works in the overwhelming majority of cases, and it can be made to work better with some constructive assistance from the Federal authorities. Nevertheless, there is a need for sanctions in instances where parties choose to ignore their obligations.

We strongly believe that sanctions should never become the tail that wags the dog; that the most effective sanctions are promptly imposed administrative sanctions, injunctive relief, and tort recovery limitation; and that civil and criminal penalties should be reserved for continuing, egregious violations of key provisions of the law.

Most States have some form of civil penalty; some have criminal penalties as well. A few (Massachusetts, Connecticut, and New Hampshire) have systems of administrative sanctions or citations.

Civil and criminal penalties, by and large, are not used, generally because such remedies are cumbersome, difficult to invoke, and require the cooperation of the State criminal justice system. Most State attorney general and district attorney offices are not interested in pursuing one-call violations unless there has been loss of life, significant personal injury, or significant property damage; they simply have too many other things to do. By the same token, they do not wish to delegate their enforcement powers to one-call systems, or facility owners and operators.

Moreover, we, as one-call systems, do not want to become enforcement agencies; it would confuse the public as to our role, distract us from what we do, and lead to fewer calls. We do see our role as providing information to parties to permit the law to be enforced by those with the authority to do so. Currently, we provide information in matters of civil litigation, and in the periodic requests for injunctive relief which are filed by facility owners and operators against excavators. We have done so, and can continue to do so in administrative, civil and criminal cases as well.

While there certainly have been and will be cases where the full force of the law must be employed, a well-designed and properly-run system of citations or other timely administrative sanctions remains the most likely to produce the results sought. The penalties do not need to be large, perhaps in the range of $100 - $500 (with $250 a good mid-point), analogous to a traffic ticket.

Most people do call, and most people do respond and locate. The ones who do not either are exempted by law from so doing, don't know of their obligation to call, or have a reason for not calling. Withdrawing exemptions will take care of the first group. Education

84-088 O - 94 - 4

generally will take care of the second group. Citations, perhaps after warning, will get the attention of most who don't want to call, and may outweigh the reason for not calling - particularly if they are progressively increased. Moreover, a State could choose to combine citations with its bidding process on State contracts, and suspend or limit the right to bid in the event of a pattern of not calling. Insurance carriers could be required to be notified, so that they might apply additional pressure by threatening cancellation.

Restricting tort law recovery by persons whose failure to notify or to participate has caused damage, is another effective means of compelling compliance. This system exists in Pennsylvania, for example, and it is a potentially useful means of getting the attention of people in a creative way. It does not depend upon reference to the criminal justice system, and can be used by any excavator or facility owner in an appropriate case. It holds out both a benefit - a positive - to encourage participation and a detriment - a negative - to discourage lack of participation.

One important consideration is the Law of Unintended Consequences: any penalty system which is perceived as too severe may encourage persons who have not called to cover up the damage they've caused. That is perhaps the single most dangerous of all situations involving underground facilities. Any Federal or State scheme must recognize and address this fact by encouraging reporting, and by penalizing the failure to do so.

With respect to the penalty provisions of HR 4349, there is an overemphasis on penalties that could dilute the focus on education. Further, the enforcement provisions do not give the States sufficient flexibility to design, implement and experiment with their own systems. The bill should be changed to require generally that the States create systems which provide for a mix of administrative, civil, and criminal penalties, plus injunctive relief upon application of an aggrieved party.

We have additional specific comments on various sections of the bill, which are provided in the attachment to this testimony. As we mentioned at the start, this bill, with some modest modifications, has the potential to be of real assistance to improving the process of damage prevention. We commend Mr. Pallone for taking a leadership role in Congress in this endeavor, and in joining with those of us who have dedicated a substantial portion of our professional lives to making one-call work for the public.

ATTACHMENT TO
TESTIMONY OF WILLIAM P. BOSWELL AND WILLIAM G. KIGER
CONCERNING HR 4394
ENERGY & POWER SUBCOMMITTEE
HOUSE ENERGY & COMMERCE COMMITTEE
JUNE 23, 1994

PROPOSED CHANGES TO BILL AS INTRODUCED

Title — *Change to read*: To provide for the establishment of mandatory statewide comprehensive one-call systems to protect natural gas and hazardous liquid pipelines and other underground facilities from being damaged by any excavation, and for other purposes.

Purpose of Change: *Clarification.*

Section 3. — Re subsection 2. *Add* the word "mechanized."

Purpose of Change: *The vast amount of damage is done by mechanized equipment, and it has the potential to do the most harm.* This is the area in which efforts must be concentrated, and to expand the definition to include non-mechanized equipment will likely lead to confusion and a loss of focus. We have instances where people have called before planting spring flowers, and before burying a family pet (a parakeet). While admirable, it takes time and resources away from more pressing situations. Permit the States to impose the broader requirement, but do not mandate it at the Federal level.

Section 3. — *Add* subsection 6: "one-call notification system" means a clearinghouse for excavators and designers and others to notify facility owners or operators of pending work activity that could involve conflict with such facility(ies).

Purpose of Change: *Define what a one-call system is.*

Section 4. — *Change to read*: Within one year of the enactment of this Act, the Federal Communications Commission shall establish nationwide dedicated telephone numbers to be used by those wishing to contact one-call systems established pursuant to this Act.

Purpose of Change: *Do not get ahead of technology; permit staged development of a single* *number, with emphasis on single number for each State, and a national information number, as first steps.*

Section 5. *Each State shall, within three years after the date of enactment of this Act, provide for a comprehensive one-call notification system . . . .*

Purpose of Change: *Virtually all one-call systems in existence are not-for-profit corporations; they are not State-operated or State entities.* While a State should be permitted to choose to operate its own system, it should not be required to do so, and Congress should not create a situation where the existing one-call systems would have to be disestablished. According to the drafters, such a result was not intended, and this change would more properly reflect the intent.

Section 6. *Change* "established" to "provided for."

Purpose of Change: *See Sec. 5, above.*

Section 6. (2) *Delete* "containing underground facilities"

Purpose of Change: *All areas of the state should be covered.* Whether there are underground facilities is a matter to be determined after the call is made, not before. This could lead to confusion and accidents.

Section 6. (3) *Delete* "except as provided by this Act"

Purpose of Change: *This does not appear to be necessary, as the Act covers all underground facility operators, excluding only personal facilities on private property which do not serve anyone but the property owner.*

Section 6. (6) *Add* "(d) other pertinent information necessary to provide location of facilities"

Purpose of Change: *Most one-call systems require more of the excavator than the information listed in (a) - (c), and there ought to be a provision for this.*

Section 6. *Strike* subsection 8.

Purpose of Change: *One-call systems have as their principal purpose the receipt and transmittal of information regarding the time, type, and location of excavation, the "if and when" so to speak.* Requiring the systems' telephone operators to tell an excavator "how" to excavate is not only well beyond the scope of their expertise, but would be time-consuming and interfere with the systems' ability to rapidly respond to calls. If some procedural requirement is essential, we suggest the following language: "(8) provide notice to excavators of the existence of any special procedures the State has established for excavations near underground facilities." In the legislative report state that what is intended is that the one-call systems simply be required to tell the caller that special procedures exist (if they do), and that specifics may be obtained from the State or the facility owner/operator, as the case may be.

Section 6. *Strike* subsection 11.

Purpose of Change: *As noted above, the one-call system itself does not provide penalties, the State scheme does so.* Also, this is provided for in Section 7.

Section 6. Re subsection 12. *Consider* tying record retention to the State statute(s) of limitation.

Purpose of Change: *A seven-year record retention does create uniformity, but most one-call statutes regarding records retention are tied to the State contract or tort statute(s) of limitations at present.* This is not a critical issue but should be discussed.

Section 6. *Strike* subsection 13.

Purpose of Change: *One-call systems receive and convey information concerning excavation.* They do not locate unknown facilities. To force them to do so takes resources away from their prime mission, and raises liability questions. By the same token, the provision could be rewritten to permit a one-call system to receive information concerning unknown facilities discovered during the course of excavation.

Section 6. Re subsection (15), *change* "establish" to "provide for" and do not limit fee imposition to facility operators.

Purpose of Change: *Since the States do not run the one-call systems, they do not need to be required to establish the fees to be charged, or to be required to limit those upon whom the fees are charged.* While it has been, to some extent, an article of faith that callers should not be charged for making the call, there are situations where excavators contribute to the costs of operating the one-call system. Permit the States to set limits, if they wish, but do not require it. Not all States restrict fee collection to facility operators, although this is the case in most States, and facility operators pay the overwhelming majority of all costs. Permit the States to have the flexibility to determine when fees should be recovered from others, and the use to which those fees may be put. For example, in Pennsylvania contractors (generally through their associations) pay small annual fees which are used to offset the cost of participation by municipal members, and for technology enhancement. They do so willingly, and even insisted that such a provision be a part of Pennsylvania's one-call law.

Section 6. *Strike* subsection 16.

Purpose of Change: *We understand that the citizen suits provision was intended to refer to facility operators and excavators and not to the general public.* If this is the case, the language ought to be rewritten to require that the States include in their laws provisions to permit injunctive relief to be obtained in cases where a party fails to comply with the notice or participation provisions of the State law. Although not wishing to be placed in the position of suing a State to enforce its compliance (the possible loss of highway funds should take care of this), excavators and facility operators are perfectly willing to pursue all others who don't participate or who don't provide notice.
This appears to be something which could adequately be covered under Section 7 (d).

Section 7. In general, *change* to a requirement that each State have a variety of enforcement mechanisms, including administrative sanctions, injunctive relief, civil fines, and criminal penalties. Permit the States to set appropriate limits on penalties and the conditions of their use. Eliminate subsection (b). Change subsection (d) to read "Each State shall permit one-call systems and facility operators and excavators to obtain appropriate injunctive relief." Eliminate subsections (e) and (f).

Purpose of Change: *One-call systems do not set or enforce penalties, issue injunctions, or revoke licenses: the States themselves do so.* Each State should be required to have a mix of penalties, which it can determine how best to employ. The Secretary of the Federal Department concerned could set up pilot programs to assist in ascertaining which enforcement mechanisms are most effective, and then recommend them to those States which do not make use of them. Since the State scheme already will permit such penalties, it could implement appropriate recommendations with relative ease.

Section 7. This section should be *reworked* to provide that the States shall have a range of sanctions for non-compliance with their statutes. It is too detailed and too focused on large civil and criminal penalties, and may have the effect of causing excavators who have not called to conceal damage rather than report it, if they believe that they might face a huge fine.

The range of sanctions should begin at warnings and citations, including administrative fines of $100 - $500 and perhaps "points" which could lead to suspension of an excavator's right to bid on State and local contracts. It should include provisions for injunctive relief upon petition of an owner/operator, or the State. It should include provisions for civil penalties and criminal action, but reserve them for the more egregious violators, when other actions have failed to produce results, and where there has been personal injury or serious property damage.
We suggest that the amount of the maximum fines be left to the States, but if a maximum is considered necessary in this statute, it be limited to $5,000 per violation and perhaps $250,000 per series of violations. It seems to us that a $5,000 maximum is not so high that a violator would likely risk concealing his violation to avoid the possibility of being assessed it, whereas $25,000 may lead one to seriously consider doing so.

Subsection (b) should be stricken. Federal and State criminal and environmental laws already permit the States to impose harsher penalties in these types of cases. This is an unnecessary duplication of existing law, which will be enforced under these circumstances.
Subsections (c) - (f) should be amended to reflect that it is the State which deals with these matters and not the one-call system itself. Subsection (e) should be stricken, as it presupposes that States routinely issue licenses to do business, or that excavators are licensed in the first place. Requiring that the States establish such schemes would be a monumental undertaking, would invite substantial opposition from the States and from those who are not now licensed, and would not add measurably to the benefits envisioned by this Act.

Purpose of Change: *Avoid micromanagement.* States are quite capable of running this part of the business, and should be given the flexibility to do so.

Section 8. *Add* a new subsection (b) "A State may delegate the requirement to compile data on operations and effectiveness to the one-call system(s) within its jurisdiction."

Purpose of Change: *Since the one-call systems have the data, they are perhaps the best collection points.* The data compiled by the one-call system(s) could be forwarded to the State to send to the Federal authorities.

Section 10. *Change* subsection 1 ("establishing") to read "providing for." Change subsection 7 ("the extent to which all underground facilities participate") to read "the extent to which all underground facility operators participate."

Purpose of Change: *Self-explanatory.*

Section 10. *A new subsection (b) should be added* to permit the States to gather the necessary information through the one-call systems. Also, the information gathered should be sent to the Secretary of the Federal Department concerned, and not to Congress. The information sought ought to be simple to obtain and to report; subsections (3) and (4) will not meet that test and should be stricken from the list.

Purpose of Change: *Simplify reporting procedure.*

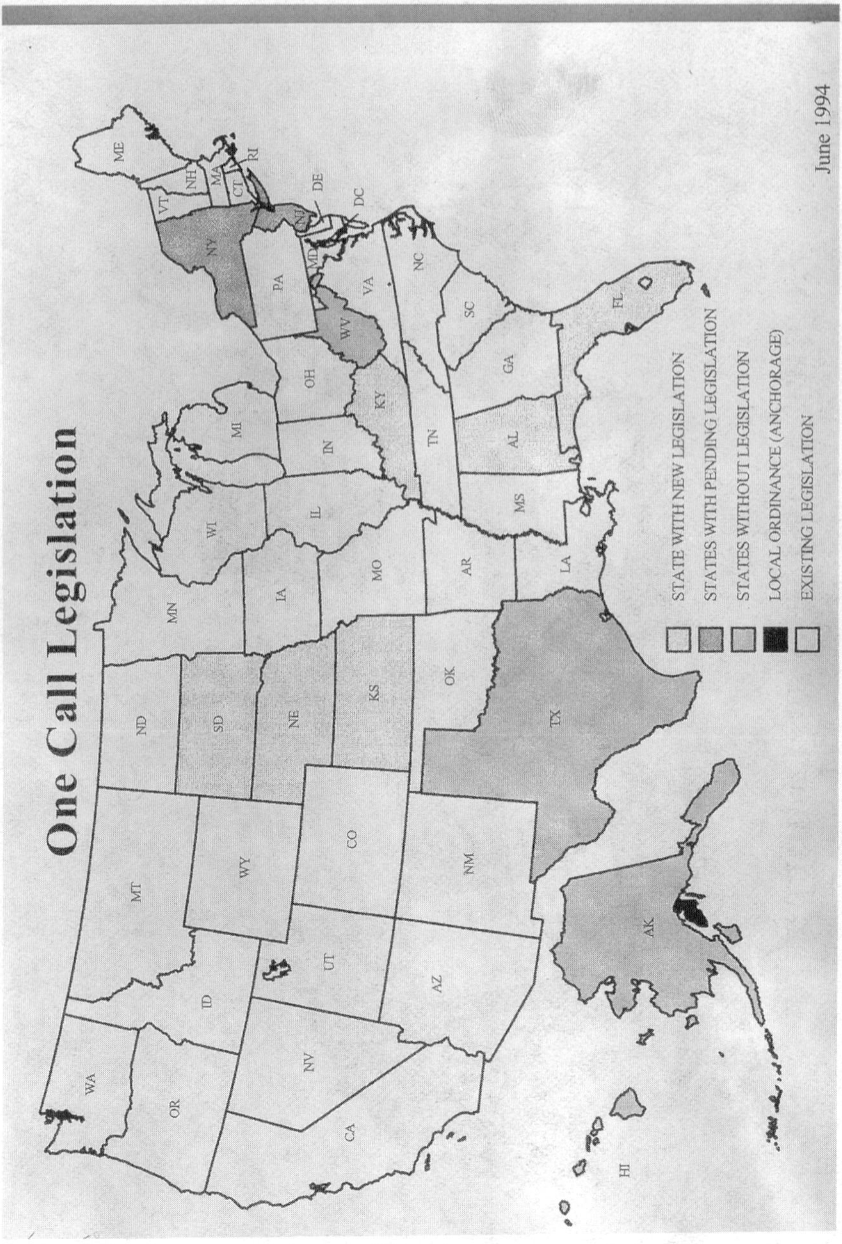
One Call Legislation
STATE WITH NEW LEGISLATION
STATES WITH PENDING LEGISLATION
STATES WITHOUT LEGISLATION
LOCAL ORDINANCE (ANCHORAGE)
EXISTING LEGISLATION
June 1994
ME
NH
VT
MA
CT
RI
NY
NJ
DE
DC
MD
PA
VA
NC
WV
SC
FL
GA
OH
KY
MI
IN
TN
AL
IL
WI
MS
MO
AR
LA
MN
IA
ND
SD
NE
KS
OK
TX
MT
WY
CO
NM
ID
UT
AZ
WA
OR
NV
CA
AK
HI

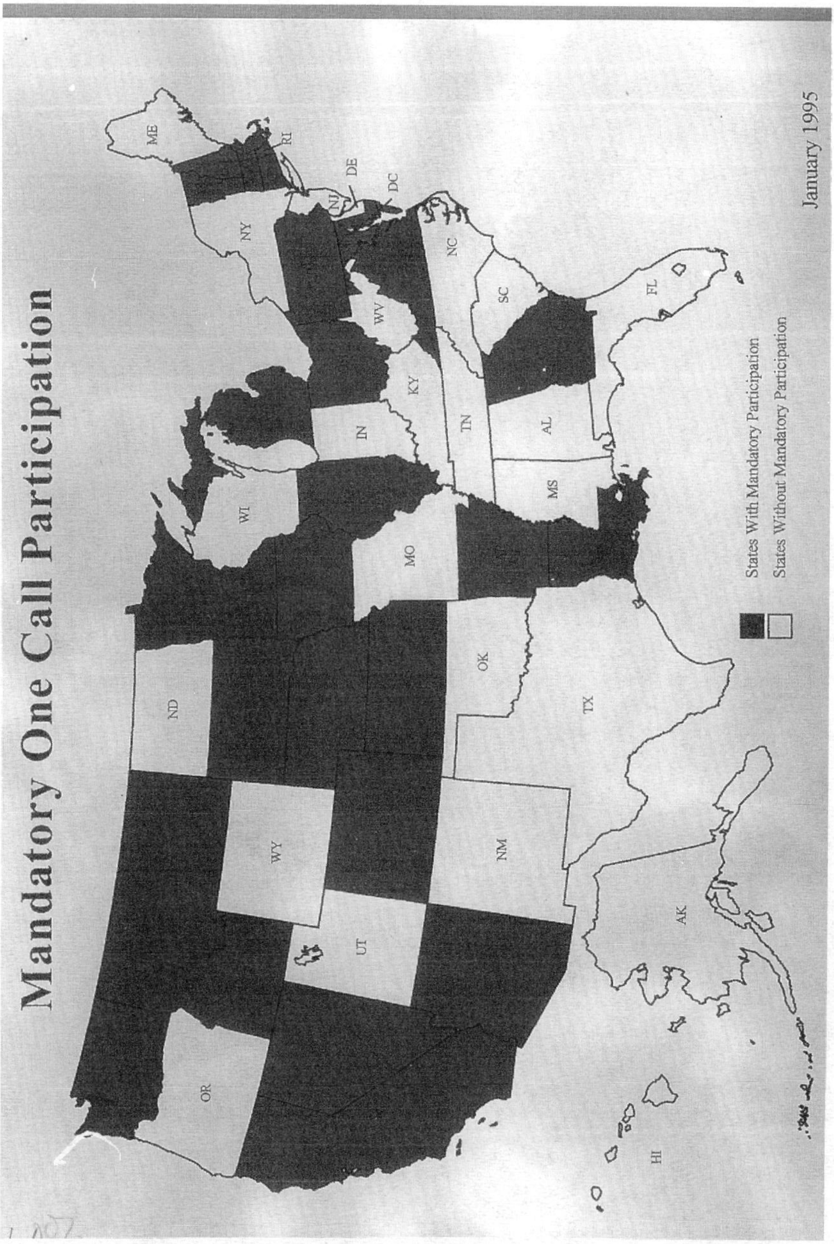
Mandatory One Call Participation
ME
RI
NY
NJ
DE
DC
WV
NC
SC
FL
KY
IN
TN
AL
WI
MS
MO
OK
ND
TX
WY
NM
UT
AK
OR
HI
States With Mandatory Participation
States Without Mandatory Participation
January 1995

One Call Centers by Type

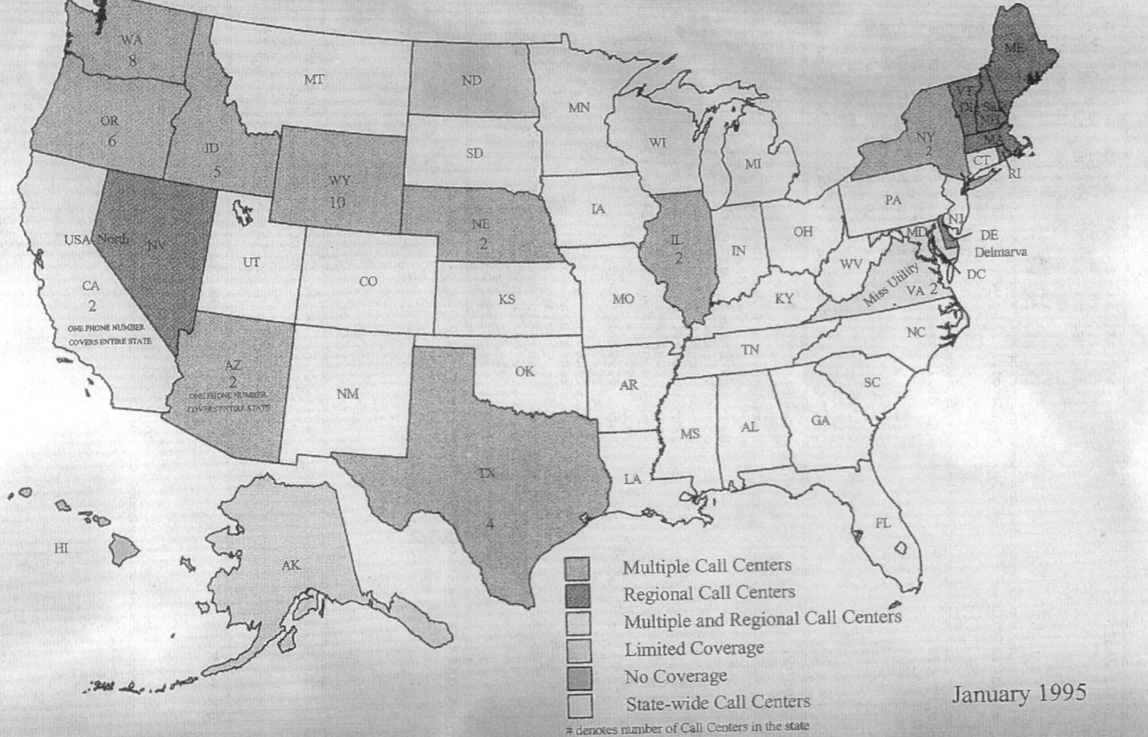

1994
S U M M A R Y

LEGISLATION:	47	Yes - AL, AZ, AR, CA, CO, CT, DE, D.C., FL, GA, ID, IL, IN, IA, KS, KY, LA, ME, MD, MA, MI, MN, MS, MO, MT, NE, NV, NH, NJ, NM, NY, NC, ND, OH, OK, OR, PA, RI, SC, SD, TN, UT, VT, VA, WA, WI, WY
	4	No AK, HI, TX, WV
PARTICIPATION:	28	Mandatory - AZ, AR, CA, CO, CT, D.C., GA, ID, IL, IA, KS, LA, ME, MD, MA, MI, MN, MT, NB, NV, NH, OH, PA, RI, VT, VA, WA
PROTECTS:	42	All Utilities - AL, AZ, AR, CA, CO, CT, DE, D.C., NM, MN, WY, GA, ID, IL, KS, KY, ME, MD, MA, MI, MN, MS, MO, NV, NH, NM, NY, NC, ND, OH, OK, OR, PA, RI, SC, SD, TN, UT, VT, VA, WA, WI, WY
	2	Gas Only - FL, NJ
	1	All Owners - LA
	1	Utilities in Public and Private right of ways - MT
EXEMPTS:	12	Agricultural
	10	Homeowners
	3	Commonwealth & Agencies
	7	D.O.T. - AR, CA, GA, NV, NJ, PA, SD
	1	Non-pressurized sanitation & storm sewers
	2	State & Local Government
	2	Excavation less than 15"
	6	Extraction of natural resources
	5	Normal road maintenance
	1	Replacement of highway guardrails
	7	Railroads
	1	Rural Electric Coop
	1	Cable TV
	4	Conditional
	3	Non-powered hand tools
	3	Grave in cemetary
	4	Landscaping/Gardening less than 12"
	1	Emergency for lines & streets
	2	Power poles
	1	Private septic systems
ACCURACY OF LOCATIONS:	1	@ 12" - NM
	18	@ 18" - AL,AR,CO,CT,D.C.,IL,IA,KY,LA,MI,MN,NE,NC, OH,PA,RI,SC,WY
	15	@ 24" - AK,AZ,CA,FL,GA,ID,KS,NY,OK,OR,TN,UT,VA,WA, WV
	1	@ 30" - NV
	2	@ 36" - ME,MD,
	1	@ 18"Horizontal - 24"Vertical - IN
POSITIVE RESPONSE:	32	Yes - AZ,AR,CA,CO,CT,DE,D.C.,FL,ID,IA,KS,MD,MA, MI,MS,MO,MT,NE,NV,NJ,NY,NC,ND,OK,OR,PA,SC,SD, TN,VA,WI,WY
	19	No - AL,AK,GA,HI,IL,IN,LA,ME,MN,NH,NM,OH,RI,TX,UT, VT,WA,WV,KY
COLOR CODE:	46	Yes - AL,AK,AZ,AR,CA,CO,CT,DE,FL,GA,ID,IL,IN,IA,KS, KY,LA,ME,MD,MA,MI,MN,MS,MO,MT,NE,NV,NJ,NM,NY, NC,OH,OK,OR,PA,RI,SC,TN,TX,UT,VT,VA,WA,WV,WI, WY
	5	No - D.C.,HI,NH,ND,SD,

PENALTY CLAUSE:	26	Fines - Minimum $50, Maximum $10,000 - AZ,AR,CA, CT,DE,D.C.,GA,ID,IL,ME,MD,MA,MI,MS NV,NH,NM,NY,PA,RI,SC,TN,UT,VT,WA,WI
	2	Jail - MI,PA
	5	Liable damages - KY,MN,MT,TN,WA
	3	Injunctive relief - AZ,CO,MI
	1	Suspension of business - MO
EMERGENCY CLAUSE:	45	Yes - AL,AK,AZ,AR,CA,CO,CT,DE,D.C.,FL,GA,ID,IN,IA, KS,KY,LA,ME,MD,MA,MI,MN,MI,MO,MT,NE,NV,NH,NJ, NM,NY,NC,OH,OK,OR,PA,RI,SC,SD,TN,UT,VA,WA,WI, WY
	6	No - HI,IL,ND,TX,VT,WV
ISSUE PERMITS:	8	Yes - AK, CA, D.C., ID, KY, ME, MA, PA, WY
	43	No - AL,AZ,AR,CO,CT,DE,FL,GA,HI,IL,IN,IA,KS,LA, MD,MI,MN,MS,MO,MT,NE,NV,NH,NJ,NM,NY,NC,ND,OH, OK,OR,RI,SC,SD,TN,TX,UT,VT,VA,WA,WV,WI
NOTIFICATION TIME:	23	@ Min. 2 days - LA,CO,WY,MO,SD,UT,CT,FL,D.C.,DE,NY, OK,OR,VT,ME,VA,IL,MN,RI,MD,AK,OH,NC
	14	@ Min. 2 Working days - KY,NV,AZ,NM,IN,IA,MS,MT,WA, AL,ID,NE,AR,CA
	7	@ Min. 3 days - ND,WI,SC,NJ,TN,GA,NH
	5	@ Min. 3 Working days - KS,WV,MI,PA,MA
STATEWIDE CENTERS:	42 -	AL,AK,AR,CO,CT,DE,FL,GA,IN,IA,KS,KY,LA,ME,MD, MA,MI,MN,MS,MO,MT,NE,NV,NH,NJ,NM,NC,ND,OH,OK, PA,RI,SC,SD,TN,TX,UT,VT,VA,WV,WI,WY
CENTERS OPEN 24 HOURS:	18 -	OH,IL,MA,MI,MN,MO,PA,TX,WV,OR,WA,ID,ME,RI,VT, WY,NH,IA
MULTIPLE CENTERS:	12 -	AZ,CA,ID,IL,MD,NE,NW,OR,TX,VA,WA,WY
REGIONAL CENTERS:	4 -	ME,MA,RI,NH,VT - Nonprofit CA,NV - Nonprofit D.C.,VA,MD - Contract MD,DE,VA - Contract

TOTAL CENTERS: 75
STATEWIDE CENTERS: 42
STATES WITH NO CENTER: 1 - HI

CENTER TYPE: 19 - Nonprofit
46 - Contract
10 - Member Utility

*Effective 01/01/95

Mr. SHARP. Thank you very much, Mr. Boswell. The Chair now recognizes himself for 5 minutes and the first question really is to you, Mr. Boswell. You indicated the first principle and goal was mandatory membership and earlier in your testimony you had identified the fact that in I believe all but Hawaii, some form of one-call exists. Can you give us any data with respect to how many of the States actually have mandatory membership as the major principle?

Mr. BOSWELL. Twenty-seven of the 49 States. By the end of this year, that will be 2 more States, so that will bring it up to 29; 27 of the 49 States have got mandatory membership right now. The trouble is, there are some exclusions to mandatory membership in a bunch of the States——

Mr. SHARP. That is not universal, in other words.

Mr. BOSWELL. Yes, sir.

Mr. SHARP. It is mandatory membership but only for certain aspects.

Mr. BOSWELL. Normally the people exempted are governmental authorities. Unfortunately, that doesn't always work because a lot of governmental agencies engage in a lot of construction and own a lot of facilities and we think they ought to be included in this.

Mr. SHARP. Anybody else want to comment on that question?

I mean, that is obviously one of the chief concerns that I think drives Mr. Pallone and many of us is that we don't have universal participation in the systems that exist. Let me ask second, and some of you addressed this issue to some degree and it is going to be the most complicated one I think we have to deal with, and that is the question of the Federal, State relationship in this. Most of the proposals are designed to try to see to it that it is simply a State-run program and to get some minimal guidelines for all States.

The question is to what degree do we use the Federal power to accomplish that and what techniques? And I don't know if any of you can give us a suggestion on the approach you think will best take on this. The idea of cutting off all their highway funds or something of that sort which has been proposed does not always set well, we have noticed, and does not seem to inspire consensus in order to accomplish the goal. Mr. McMahan.

Mr. MCMAHAN. Mr. Chairman, I will speak to the concept of uniformity and having a State-wide system, because many of the systems that exist today are very local in nature. They aren't comprehensive State-wide or across State lines.

Most of our pipelines do cross State lines and in having a—and in many contractors, most contractors work in many States. So having a system that is consistent, the guidelines are standard across the State, facilitates excavators in being able to make the contact when they are going to be doing some excavation work.

Clearly companies that work across State lines and have facilities across State lines, a consistent set of guidelines put out to the States, again, which would enable them to incorporate existing systems and what we have learned to date in their programs. They can draw on the experience that is there to date but a consistent set of guidelines would be very beneficial to—I feel to the operator, as well as the excavator.

Mr. SHARP. Mr. Riordan, I wasn't clear in your testimony. You do favor the Federal Government requiring this of all States or you favor us providing some incentive to all States, which almost always is a financial incentive if it is a true carrot.

Mr. RIORDAN. Mr. Chairman, we favor that the Federal Government sets guidelines in terms of incentives. That would be a secondary approach to it.

Mr. SHARP. You said "guidelines." In other words, we outline what we think would be a good program and it is strictly up to the State in whether or not they should engage in that program? Is that what you mean?

Mr. RIORDAN. We have not come up with a way of incentivizing, if there is such a word, the States or penalizing them. But we think there has to be very firm mandates for all of the States in terms of one-call legislation and those guidelines at some time need to work, so exactly how we do that, we have not come up with a program.

Mr. SHARP. Mr. Boswell, I wasn't clear whether you addressed that point. My impression was that you thought we might create a model but then leave it to the States, or did you have an advocacy position on that issue?

Mr. BOSWELL. I guess no lawyer is capable of giving a short answer, but let me take a stab at it. We think there ought to be some guidelines that the Federal Government establishes and we think the guidelines ought to be in the form of mandates to the States within the broad general guidelines.

If you look at our prepared written testimony, you will see that I have urged the committee to stay away from crossing T's and dotting I's because I think that is best left to the States which have been doing it for decades now really. We appreciate the Federal Government's role in this to the extent that there are certain things which we have identified within the one-call industry over the years which we believe make a good one-call system, and a lot of contractors and a lot of our members do participate across State lines.

To the extent that there is a certain degree of uniformity in that, it makes life easier for everybody, so we do think that there ought to be some mandates. We do think that some of the specifics ought to remain within the proper jurisdiction of the States because each State may have a different set of circumstances which it considers important as a matter of public policy and we think the States ought to be left the freedom to experiment.

There is nothing which says that something that comes out of here is any better necessarily than what the States have done. We suggested that the DOT get involved to the extent of producing a model program.

One-call Systems International has a model statute which we have given to the States for years now, and many of them have modeled their laws after that model statute. I think if it came from Washington, that would give it a little bit—I hope it would give it a little bit more authority, if you will.

Mr. SHARP. You think, huh?

Mr. BOSWELL. Hope springs eternal.

Mr. SHARP. Well, I appreciate that and I want to recognize my colleague after just indicating that obviously we want to be careful that we are not adding to the burdens that State governments face in terms of our putting a mandate on for which they then have to raise taxes in order to fulfill the mandate. They clearly are not in a mood to have us do that kind of thing and there will always be an argument over what constitutes taxes, but the systems that generally are in place provide—can fairly easily provide a mechanism for funding by the user fee system, and we need to discuss obviously with the governors and the State associations, State governments, because we want to be sensitive to that proposition as we try to move forward in how to elicit their cooperation as best we can.

Yes, Mr. Boswell.

Mr. BOSWELL. Mr. Chairman, I would suggest that from our perspective, we don't see a major or even a very significant financial burden on the States themselves. Remember that the systems are not State-run systems and I agree with Mr. Riordan, they should not be State-run systems, they don't need to be because the ones which exist right now are not. And we are essentially self-funding.

The initial draft of the proposal from Mr. Pallone indicated that the fees should be recovered from the facility operators and we would suggest that you don't need to make it quite that narrow and I think your draft does not. Let the States have the authority to decide where the fees come from. The facility operators, by and large, together with the contractors who are very big supporters of one-call systems, have been willing to fund the cost of this effort. Most people who call pay nothing at all.

In Pennsylvania, my State, for example, the contractors pay. Interestingly enough, they pay because they insisted upon it in the latest draft of our legislation. Originally we wanted no one to pay because we thought it was intuitively correct that if you force people to pay to make a phone call, they won't make a phone call. Much to our surprise, the contractors got together with the municipalities and as a means of funding the cost of municipal membership, they said, we will pay an annual fee to belong to a one-call system and they mandated that as a matter of legislation at the State level and it is working just fine and the moneys that we raise go to offset the cost of municipal membership and for technological enhancement.

Mr. SHARP. Thank you very much. I want to recognize my colleague from Illinois and at the conclusion of his questions, we will have to take a break for us to vote in the House of Representatives.

Mr. HASTERT. I thank the chairman. One of the things we kind of talk about here is pipelines and what is the definition of a pipeline. I am not sure that is clear. I think when you talk about municipalities delivering natural gas to homes, is that a pipeline, is that a municipal delivery system? Are they included? Aren't they included under this?

And yet I think an 8-inch main is probably just as dangerous as a 20-inch main or whatever size natural gas mains are, and so I think we need to have some further scrutiny on that as well, and when you start to say—in Mr. Pallone's legislation, you are charging almost $100 a mile, that would be, I would think, when you look at all the other participants in that, water, municipal water distribution areas, electrical lines, cable TV, telephone lines, all subterranean, I am not sure if we are putting only that onus on natural gas pipelines or on all, and of course they all share that risk, especially I guess gas and electricity are probably the most dangerous if you hit into those things, but also a great public risk are telephone lines and those things as well, and I know in my State are all a part of the Julie system.

One of the things I think we also need to define a little bit are guidelines and standards. Mr. Riordan, you talked about guidelines. Maybe we need standards.

Mr. RIORDAN. Yes, I think—I think standards are probably more appropriate, but there may be—I think there needs to be some direction given, but I think we will end up setting standards. I might comment that I think what we are trying to do here is—one of the problems with it, it makes good common sense, what we are trying to do and what we want to do is come up with a common sense solution to it, so I think that standards would be appropriate.

Mr. HASTERT. That is a pretty important word, common sense. That is not used enough. But anyway——

Mr. RIORDAN. That is why we are all pretty much agreeing on what we are doing here. That is kind of unusual.

Mr. HASTERT. Mr. McMahan, there is a statement made that remote control valves must not interfere with inspection, and the remote control and automatic valves, is there a danger in saying, mandating these have to be in place and what—I mean, are they, for instance—a screwed on valve is one thing. But what jeopardies are there with remote control and electronically-controlled valves and those types of things?

Mr. MCMAHAN. They are very different animals when we talk about automatic valves versus remote control valves. Certainly on the oil side of the industry, we have supported having more remotely controlled valves to reduce the outfall from any kind of a release, and being able to do that from some control center somewhere.

Having valves that react automatically have been a problem for the industry in that if they react at the wrong time automatically to shut the—a pipeline off, it can create problems in itself, overpressuring the pipeline and that sort of thing, but certainly the application of remotely controlled valves does have a——

Mr. HASTERT. Is there a problem of getting pigs through these valves as opposed to manually controlled valves?

Mr. MCMAHAN. No, not really. It is really how you control the opening and closing of the valve more so than the innards of the valve so to speak.

Mr. HASTERT. Mr. Hall, you talked about—we actually talked about FERC setting pipeline routes in urban areas, and you have deferred to that, I think probably wisely, but Mr. Hall, you were talking about siting pipelines in urban areas. Do you think that is pretty much a municipal responsibility or does Department of Transportation want to do that? I mean, not only pipelines, but you have other interconnect types of subterranean.

Mr. HALL. Well, and under present regulation of course, FERC has that responsibility and those resources are available in its agency and we consider FERC a partner in terms of safety and in terms of siting of these interstate pipelines.

Mr. HASTERT. Who is actually doing the siting today? I mean, you are talking about interstate, but I am talking about if—you know, Northern Illinois Gas in my district wants to run a pipeline into a subdivision, is that the subdivider or is that the State or is it the local municipality or is it your organization? Who makes that determination?

Mr. HALL. I would assume in that situation, sir, it is usually the local jurisdiction that has control.

Mr. HASTERT. But you don't know that for sure?

Mr. HALL. No, sir.

Mr. HASTERT. You might want to get that information to us.

Ms. MOLER. We regulate interstate pipelines. States regulate intrastate pipelines. They give a franchise to local distribution companies and then they have authority to hook up homes and the like.

Mr. HASTERT. The State gives the franchise?

Ms. MOLER. The State gives the franchise to the local utility.

Mr. HASTERT. Like the Interstate Commerce Commission or the Public Utilities Commission?

Ms. MOLER. It varies from State to State. Some of them have State siting commissions and some of them have State public utility commissions.

Mr. HASTERT. Is there a definition of pipelines and——

Mr. SHARP. We have about 4 minutes left to vote. If the gentleman has further questions.

Mr. HASTERT. That is all I have. I would just like to have the response to that. You don't need to do it now. If you would do it for the record, I would appreciate it.

Mr. SHARP. Ladies and gentlemen, we appreciate your time and attention to this and we will let you go and we will call up the next panel as soon as we return from this vote. Thank you very much.

[Brief recess.]

[Testimony resumes on p. 127.]

[The following responses were received:]

Questions Submitted to the
National Transportation Safety Board by
Chairman Philip R. Sharp

Question 1: Do you agree or disagree with Chair Moler's assessment of siting provisions referred to in H.R. 4616? Do you think that FERC should have a greater role in considering safety when siting pipelines?

Answer: Chair Moler indicated in her April 19, 1994, testimony before the Senate's Committee on Energy and Natural Resources that the Federal Energy Regulatory Commission (FERC) has the following responsibilities under the Natural Gas Act:

> To make a "determination that the facilities are in the public interest and should be approved" and for "siting of the pipeline." Later in her remarks, Chair Moler added that "Above all, the Natural Gas Act requires us to analyze whether a project is in the "public interest. This means we focus on land-use issues. We try to avoid or minimize the impact on local land-use issues. We try to avoid or minimize the impact on residences, schools, hospitals, and business ...."

The U.S. Department of Transportation's (DOT) Office of Pipeline Safety (OPS) does not have similar authority. Also, state and local governments have no authority to control interstate pipelines. Therefore, the Safety Board believes it incumbent upon the FERC that it consider the siting and the consequential affects of potential pipeline accidents when making determinations as to the "public interest."

While DOT can write design standards, it cannot consider the effects a pipeline may have on alternative routes and the relative exposure due to development, topography and other factors. Chair Moler stated in her testimony before the House Subcommittee on Energy and Power, Committee on Energy and Commerce, on June 23, 1994, that they conduct site visits during review of alternate routings and they consider public comments, "one of the best ways to identify both potential problems and possible alternatives."

All pipelines subject to OPS jurisdiction must comply with OPS safety standards whether or not specified by FERC as a part of its certification. Even so, pipeline accidents continue to occur and when located in urban areas, there is a greater likelihood that they will experience failures, especially excavation damage caused failures. Consequently, the siting of and required rights-of-way widths for new pipelines in or near urban areas are essential parameters that must be considered in determining whether a proposed project is in the interest of the public in the area of the proposed pipeline. The Safety Board believes that FERC should require risk assessments as a part of its certification requirements to assist it judging whether a particular project is in the "public interest." Such assessments should address the potential consequences of a rupture explosion or radiant heat from a major fire involving a rupture of the proposed pipeline to adjacent property and people.

Question 2: Should there be exceptions to universal, mandatory participation in a

one-call system? Do you think that exempting gathering lines and flow lines undermines our goal of universal, mandatory participation?

Answer: The Safety Board believes that all operators of buried facilities and all excavators need to participate in an excavation damage prevention program for it to achieve maximum effectiveness. The only exception we believe might be made without adversely affecting the safety objective of an excavation damage prevention program is where the only buried facilities located on land controlled by the operator of the facilities and which only serve the operator. Such an exception would exclude facilities like as large industrial complexes and gathering fields where the operator has control over all activities, including excavation.

The Safety Board believes that an exclusion should be provided only where it can be demonstrated that the exclusion will not adversely affect public safety. Exclusions should not be provided to gathering pipelines systems as an industry as many such pipelines are on lands not under the full control of the operator and are located near residential and commercial buildings. Many gathering pipelines, from a public safety viewpoint, pose threats to public safety much like natural gas and liquid transmission pipelines. They often operate at similar pressures, transport similar or more hazardous materials, and are not required to comply with the Office of Pipeline Safety's standards.

The Safety Board reviewed the 1992 RSPA accident reports for accidents involving gathering lines, even though those in rural areas are not required to be reported. Two accidents reports were found that involved excavation damage. The Safety Board also reviewed its records of investigations and identified 15 accidents that involved gathering lines since 1975. Three of the 15 accidents involved excavation and another accident involved improper construction.

RSPA reviewed the accidents telephonically reported to the National Response Centers and determined that between 70 to 165 notifications that include the word "gathering" have been received annually since 1987. Data the Safety Board has recently compiled on accidents reported to RSPA indicate that about 21 to 22 percent of all accidents involve excavation accidents. Thus perhaps there are as many as 15 to 35 accidents annually that involve gathering lines and excavation damage.

The Safety Board currently is investigating a May 3, 1994, accident involving a 10-inch diameter natural gas gathering pipeline. The pipeline was ruptured near Green River, Wyoming, when excavation equipment operated by a highway contractor struck the pipeline. Two contractor employees were killed immediately and two others (one later died) were injured. The contractor was performing work for the Wyoming Department of Transportation and that State's law did not require all operators of buried facilities to participate in available one-call notification systems. There was a one-call system operating in the area in which the rupture occurred, but the pipeline operator did not participate.

Question 3: What are the essential elements that should be included in Federal legislation as minimum requirements of a one-call system? (Please refer to my discussion draft of the amendment in the nature of a substitute to H.R. 4394.)

Answer: Based on our preliminary discussions with the representatives of the States of Arizona, Connecticut, Massachusetts, Minnesota, and Utah, we believe the following elements are essential for a program to be highly successful in preventing excavation-caused infrastructure damages:

Full participation in the program by all organizations that operate buried facilities, whether public or private (as currently required in section 5(a)(1));

Full participation in the program by all organizations that excavate, whether public or private (as currently required in section 5(a)(1));

Provide a true, statewide one-call notification center and specify the conditions for its use (as currently required in section 5(a)(1)(B));

Provide for prompt, effective administrative sanctions for violations, injunctive relief against habitual violators and imminently dangerous operations, and criminal sanctions for persons whose actions violate the act and results in casualties or substantial property damage (as currently required in section 6); and

Requirements to educate the public, excavators, and buried facility operators on the program requirements and how to comply with the requirements.

The Safety Board will conduct an excavation damage prevention workshop on September 8 and 9, 1994, which will bring together individuals experienced in the operations of excavation damage prevention programs. From that workshop, we will learn if there are other elements necessary for developing effective programs and we will gain more detailed information on each element. We believe that this workshop will provide for both the OPS and the Safety Board the information needed to motivate future improvements in these programs.

In the interim, however, we offer for your consideration the following comments:

We believe that more time than needed is afforded in the legislation for states to implement its requirements. Most states already have a statewide one-call notification center and have excavation damage prevention laws that could be amended readily to comply. We believe that most states could comply with the requirements within two years. The legislation already provides a provision for the Secretary to extend the application date for a year should any state be unable to comply. (Section 4(a)).

The legislation should make clear that the states are to establish excavation damage prevention programs not one-call notification centers. All current centers have been established through voluntary, cooperative relationships among buried facility operators. Their structures and operating procedures have been developed to address local conditions unique to the organizing agencies. The Safety Board believes that flexibility has been one of the more important factors in the successes that have been achieved by existing programs. Consequently, we encourage the Committee to retain as much flexibility as possible relative to the organizational structure and the day-today operations of one-call notification centers.

The legislation should clearly distinguish the provisions applicable to one-call notification center operations and those applicable to state governments. To assist in making these distinctions, the committee should consider using the term "excavation damage prevention program" when referring to actions intended by a state government and use the term "one-call notification center" when referring to actions required by the center that receives and disseminates information on planned excavations.

There should be a requirement for the state and the notification center to provide public, buried facility operators, and excavators information to educate them on the provisions of the state damage prevention requirements and on how to use the notification center.

Finally, a requirement for the states to develop training and licensing provisions for operators of mechanized excavation equipment should be considered.

Question 4: How important is the role of public education in the use of one-call systems? Do you have any suggestions, other than the use of public service announcements which has been provided for in the Bill, on how we can best communicate the importance of this issue to the public?

Answer: Public, excavator, and buried facility operator education on the requirements of the statute and on how to use the notification center will be a major factor in reducing excavation damage incidents, and we believe there should be requirements for educational programs without providing specifics as to how the program should be implemented. This will provide the flexibility needed for the states, the notification centers, and the operators of buried facilities to implement and then later adjust their programs as necessary to obtain maximum cooperation.

Today's more successful excavation damage prevention programs use a wide variety of means to educate people on using the notification center. Many training opportunities are in use today, including:

Requiring non-complying parties to attend training classes that detail the program requirements and that provide training to convince those attending of the benefits they will receive when properly using the notification center;

Providing a cadre of speakers to discuss the program before community and business meetings;

Visiting construction sites to discuss the program with excavation equipment operators and providing them with decals, pens, etc. to remind them of the program and the telephone number to call;

Providing booths at county fairs, baseball games, and other large public gathering; and

Installing one-call information decals on rental excavation equipment.

The Safety Board believes that effective programs for educating the public are important elements of any program aimed at improving public safety. We have encouraged improving current pipeline company public education programs now incorporated in OPS requirements as they do not appear to be nearly as effective as other public education programs. Programs such as those used to motivate active public participation in improving automotive safety by using seat/harness belts and child restraints, and improving awareness at railroad grade crossings. Those public education programs need to be reviewed to determine which of the various techniques are responsible for their effectiveness and may be used to make pipeline education programs more effective.

Question 5: Do you think the penalty provisions of H.R. 4394 are too high? What type of penalty structure should be included in Federal legislation?

Answer: The Safety Board has not addressed penalty provisions in its accident reports or safety recommendations.

MIDCON CORP.
AN OCCIDENTAL PETROLEUM CORPORATION SUBSIDIARY

John F. Riordan
President and CEO

MEMO FOR THE RECORD
Due Date 8/10/94
Hearing Date 6/23/94
Received in Office 8/10/94

July 27, 1994

The Honorable Philip Sharp
Chairman
House Energy and Power Subcommittee
331 Ford House Office Building
Washington, DC 20515

attn: Judy O'Brien

Dear Mr. Chairman:

Enclosed are my responses to your written questions on H.R. 4394 which result from the June 23rd hearing. I would appreciate having these included in the record.

With all best wishes.

Sincerely,

John Riordan

cc: Michael Baly, III
Jerald V. Halvorsen

701 East 22nd Street • Lombard, Illinois 60148-5072 • (708) 691-3100

1) Do you think that exempting any gathering lines and flow lines undermines our goal of universal, mandatory participation in one-call systems? Are there any special circumstances or situations that would justify the exemption of such lines?

There are a number of circumstances where a business or industry has exclusive operational control of the property, has the only underground facilities and would be the exclusive excavator doing any work on the property. It would seem logical in these circumstances, where the person making the call would be making the call essentially to himself, not to require the person to be included in a state one-call system. In order to receive such an exemption, however, it would be important that the owner of the facility be certain that all underground facilities belong to it and to no one else and is willing to assume liability for damage to any underground facilities on this property. This provision should not pertain to homeowners.

2) The suggestion has been made to restructure this legislation towards a more incentive-based approach, rather than a mandated approach. Would you support this concept? Would you support the collection of user fees from all underground facility operators to help finance a one-call incentive program?

The establishment of inclusive one-call systems in every state is important to the natural gas pipeline industry. Each state should view the establishment of these programs as providing a strong aid to prevent damage to all underground facilities, protection to all the citizens in that state and not as an undue regulatory burden imposed on states by the federal government. It is also important that all underground facility operators recognize that one-call systems do not increase costs, but rather help to eliminate costs. Education can be provided by the Department of Transportation by holding workshops for states to inform states of the benefits of an effective one-call system and on how states can improve their one-call systems. Costs of one-call systems are currently covered by those belonging to the system. We see no reason to change that funding mechanism.

3) How important is the role of public education in the use of one-call systems? Do you have any suggestions, other than the use of public service announcements which has been provided for in the bill, on how we can best communicate the importance of this issue to the public?

An informed public always makes a system like one-call more effective. Informing the public via notices in their utility bills, by radio and television advertising and by providing a prominent place in every telephone book is very important. In addition, special efforts should be made to communicate and educate excavators. Some ways to achieve this are through contacts with their trade associations, special mailings to individual companies, providing workshops and requiring them to participate in and to use the system.

4) Do you thinks the penalty provisions in H.R. 4394 are too high? What type of penalty structure should be included in Federal legislation?

States should be encouraged to establish penalties that are administrative, civil and criminal in nature. However, many states do not want to go through the procedures necessary for either criminal or civil penalties. Consequently, some states have found administrative penalties simpler and more effective to use. Federal legislation should not tell them how strict the penalties should be or how to structure the penalties. We believe states should be required to establish penalties that encourage people to use the system, yet do not discourage people from promptly reporting if they strike an underground facility and had not called the one-call system prior to digging.

5) What are the essential elements that should be included in Federal legislation as minimum requirements of a one-call system? (Please refer to my discussion draft of the amendment in the nature of a substitute to H.R. 4394).

1. Nationwide toll-free number.

2. H.R. 4394 requires DOT, working with other appropriate agencies, such as NTSB, to develop provisions for a model one-call damage prevention program. We believe a model program should include the following elements:

   - All operators of underground facilities that are at risk of being struck by excavators should be included in the program.
   - All excavators undertaking activities that could result in the striking of underground facilities should be required to first call the one-call system (unless the excavator and the facility operator are one in the same person and no other underground facilities are threatened).
   - There should be an enforcement effort that encourages the participation of contractors in the system and encourages contractors who strike facilities to report such strikes, but punishes contractors who fail to use the system, particularly repeat offenders.

- The program should include a public education component, which DOT would compliment with its own effort.

DOT could periodically update its model program.

3. DOT should annually report on whether state programs meet the model federal guidelines, publicize the results of the report, and recommend actions that could be taken to encourage states to upgrade their programs.

Shell Pipe Line Corporation

R. C. McMahan
President

Two Shell Plaza
P. O. Box 2648
Houston, Texas 77252

July 28, 1994

The Honorable Philip R. Sharp
Chairman
Subcommittee on Energy and Power
331 Ford House Office Building
Washington, D.C. 20515-6120

Dear Chairman Sharp:

On behalf of the American Petroleum Association and the Association of Oil Pipelines, I appreciate this opportunity to respond to the questions posed in your letter dated July 15, 1994. Please note that the answers provided below are based on your <u>Discussion Draft Amendment</u> Dated July 14, 1994 offered as a Substitute to H.R. 4394.

Question: I understand EXXON would like to exempt unregulated gathering lines and flow lines from being covered by this legislation. Do your views vary from this position?

Answer: Some concern has been expressed regarding the inclusion of gathering and flow lines not regulated by the Department of Transportation under the HLPSA. A meeting has been scheduled for August 3rd at the American Petroleum Institute to attempt to reach an industry consensus on this matter. The API will report the results of this discussion to the Subcommittee on Energy and Power.

Question: The suggestion has been made to restructure this legislation towards a more incentive-based approach rather than a mandated approach. Would you support that concept? Would you support the collection of user fees from all underground facility operators to help finance a one call incentive program?

Answer: We support federal legislation which mandates the establishment of State one call programs. The legislation should establish incentives for States to create these programs in a timely fashion and ensure that effective one call legislation is enacted.

Some of the user fees currently paid to the Department of Transportation by the pipeline industry could be used to fund a one call incentive program. However, all facility operators of one call notification systems should bear their appropriate financial burden for the establishment of an one call incentive program.

Question: How important is the role of public education in the use of one call systems? Do you have any suggestions, other than the use of public service announcements, which has been provided for in the bill, on how we can best communicate the importance of this issue to the public.

Answer: Public education is the most important way of communicating the one call notification concept. Without public education/public awareness, the use of the one call system would be minimal. Public education/public awareness is carried out by both the one call system and its members. The message is communicated by radio, television, newspapers, newsletters, contractor mailings, contractor awareness programs, trade shows, fairs, safety meetings and face-to-face meetings with excavators.

The use of public service announcements on both radio and television are very important. Other ways to communicate would be through newspaper and magazine advertising, agency or one call system sponsored workshops, provision of information on the one call system at the time of issuance of excavation or building permits.

Question: Do you think the penalty provisions in H.R. 4394 are too high? What type of penalty structure should be included in Federal legislation?

Answer: There is some merit to the argument that severe penalties may cause an excavator to cover up damage rather than report the damage to the facility owner and face imposition of fines or imprisonment. The States should be allowed to set the specific penalties. The enforcement of the penalties must be consistently applied.

Question: What are the essential elements that should be included in Federal legislation as minimum requirements of a one call system? (Please refer to my discussion draft of the amendment in the nature of a substitute to H.R. 4394)

Answer: Federal legislation must focus on public safety concerns. Minimum essential elements should include:

- Participation in a one call system by underground facility owners;
- Notification by excavators;

- Operation of the one call notification system by an independent corporation consisting of at least underground facility owners;

- One notification system covering a state or large geographical area within a state (no overlapping systems);

- Twenty-four (24) hour service; and

- State enforcement program.

We hope these answers will be of use to you and the Subcommittee. Thank you for the opportunity to provide additional input on this important subject.

Sincerely,

cc: Ms. Margaret Matheson
American Petroleum Institute
1220 L Street, N.W.
Washington, D.C. 20005

Mr. Patrick Corcoran
Association of Oil Pipelines
1725 K Street, N.W., Suite 1205
Washington, D.C. 20006

July 27, 1994

Hon. Philip R. Sharp, Chairman
Subcommittee on Energy & Commerce
House of Representatives
Washington, DC 20515-6117

Dear Mr. Chairman:

I have your letter of July 15, 1994, concerning HR 4394, the "Comprehensive One-Call Notification Act of 1994." Following are my responses to the questions you posed.

1) Incentives can be very useful, especially if there is an underlying structure in place. The States ought to be required to have damage prevention programs which include one-call types of activities. Federal guidance ought to leave the States leeway regarding the specifics of their programs, but suggested model procedures would be valuable in creating uniformity in key areas.

2) There is no hard and fast answer as to the number of dollars necessary to incentivize a program, but I believe $6-8 million ought to be a good start, with no more than that in subsequent years.

3) A well-run, state-of-the-art center should cost its members about $1-2 million/year. The States themselve don't run the systems at present, and I am unaware of any State spending any money of its own on a sysyem, except to the extent that it or any of its agencies choose to be members. State support of one-call has been moral and legal support, not financial. We've never been able to calculate with precision how much the systems save, but it is easily many multiples of the cost (i.e., in excess of 10x the cost.) For example, I am familiar with an incident involving a cable TV contractor whose personnel augered through a large diameter petroleum pipeline about ten years ago. That one incident cost the affected parties approximately $40 million, which would have paid for the operation of the Pa Once Call Syatem alone for the next 16 years. We had no fatalties in Pennsylvania from 1989 to 1994.

4) Although they are stated as maximums, which is helpful,I believe the penalty provisions are unnecessarily steep, and may encourage defaulters to cover up their mistakes. States should have a variety of penalties - administrative, civil, and criminal - and use the most appropriate

84-088 O - 94 - 5

under the circumstances. They are capable of making these decisions themselves, and Federal law should simply require that the States have a market basket of penalties.

5) Public education is critical. It has gone and must continue to go far beyond "public service" advertisements. It includes mailings, face-to-face meeting with operators, contractors and excavators, working with trade associations, and advertising media, to name a few. Education is a key component of all one-call systems with which I am familiar, and one of the most important roles DOT could play is in the arena of public education - including model statute activities. One Call Syatems International would be very pleased to work together with DOT in this endeavor.

6) In my earlier testimony regarding essential elements of a one-call system, I identified four categories. They are a) mandatory participation by operators, b) mandatory calling by excavators, c) public education, and d) penalties for non-compliance. My opinion hasn't changed since then, but some slippage as discussed in 7, below, would be acceptable if that's what it takes to get a bill passed.

7) While it would be nice if there were no exemptions, realistically that may not be possible at this time. Each State law currently is different, and the systems work fairly well. You might want to let the States have some opportunity to exempt gathering lines in unpopulated areas. The key ought to be public safety; if there isn't any significant compromise to public safety, a limited exemption may not be inappropriate.

Again, thanks for giving us the opportunity to provide input. This bill remains a good idea, and one which needs to be enacted into law in some form this session. Your staff has been wonderful to work with, and, not to lay it on too thick, that seems to me to be a function of leadership at the top.

Sincerely,

W P Boswell

Legal Advisor
One Call Systems International

Mr. SHARP. The subcommittee will come to order. Actually, let me say to our witnesses, I appreciate your effort. I am going to ask that we delay just a couple minutes on this panel because Senator Lautenberg and Senator Bradley are on their way to testify. We will let them go first, so if you folks would just stay in recess for a couple minutes and wait for Senator Lautenberg, otherwise we will disrupt the proceedings, so we will just stay in recess a few minutes. Sorry about the inconvenience.

[Brief recess].

Mr. SHARP. The subcommittee will please come to order.

We are pleased to have with us Senator Lautenberg and I think Senator Bradley is on his way. So Senator Lautenberg, we will be very pleased to hear from you at this point.

STATEMENT OF HON. FRANK LAUTENBERG, A U.S. SENATOR FROM THE STATE OF NEW JERSEY

Senator LAUTENBERG. Thank you very much, Mr. Chairman.

Mr. SHARP. Senator Bradley and I are a team, but we don't always travel together. We don't want to run the risk for New Jersey that the two of us are in the same vehicle and calamity takes place.

Mr. SHARP. I know how respected all of you are on Capitol Hill and would never want that kind of a loss.

Senator LAUTENBERG. Thank you very much, Mr. Chairman, and members of the committee. I do appreciate the opportunity to appear before you today.

The hearing for me is critical because it is yet another indication that the events of March 23rd in Edison, New Jersey and the people affected by the pipeline disaster have not been forgotten. It was a terrible night and it was a terrible experience, not just for Edison, New Jersey, but for others in our very densely populated State and for others across the country who live in areas where pipelines crisscross all over the place.

They deliver an important source of energy, but when something goes awry, as we saw that night, the whole nation witnessed what took place. It was a ball of fire, shot up in the air. It was described by those who saw it as something like a butane lighter, a couple hundred feet in the air; just this huge flash of flame and explosion that took place, as well.

Americans saw that emergency on television. Frankly, they shudder and a lot of questions were raised about what is my exposure, what is my family's exposure to something like that. It was devastating to the people involved and traumatic to almost every resident of my State, which, again, is crisscrossed with similar pipelines, and I have seen not only the physical problems because the devastation was war-like, and I have seen the war before, but the psychological problems that continued afterward were very severe, and I talked to families who lost everything virtually but the clothes on their backs, and some of the buildings in this complex were just leveled to the ground.

So the emotional fallout is pretty severe. The kids and adults who replay the events of that evening each and every night before they drift into a fitful sleep, always on the alert, and we cannot allow this type of event to happen again.

The two bills being considered by the committee today, the Comprehensive One-Call Notification Act of 1994 and the Natural Gas Pipeline Safety Act of 1994 offer a clear blueprint on how to make pipeline operations in facilities all across the country much safer. I am a cosponsor of the one-call bill and sponsor of the pipeline safety bill. I respectfully request that the committee favorably consider these two bills.

The Comprehensive One-Call Notification Act of 1994 includes what I believe to be the critical elements to any effective one-call system. Those elements are mandatory participation, quick and sure justice, and effective public awareness activities. There is a clear public interest for the Federal Government to insure that States adopt strong programs which require participation by all utilities, construction companies, and State and local agencies who might endanger the integrity of the gas pipelines.

In talking to pipeline officials, they tell me that frequently those who violate the integrity of pipelines by digging or construction work around there are often government agencies, so we have to strengthen the notice process to make sure that people know what they are getting into when they start to dig in those areas.

Beyond the required compliance by parties engaged in construction, it is in the public's interest that penalties guarantee swift and sure justice for violators, and the public is aware of its responsibilities and complies. Strengthening one-call State requirements can prevent the devastation witnessed in Edison.

The Natural Gas Pipeline Improvement Act of 1994 addresses pipeline safety concerns raised by the Edison disaster which complement and expand on the comprehensive one-call bill. Since Senator Bradley and Congressman Pallone and I just introduced this legislation on Tuesday, perhaps a brief outline of the bill will help.

Our bill authorizes the DOT, U.S. DOT, to recoup the cost of accident investigations from pipeline companies so that its normal inspection program, that is the one that goes on regularly, is not interrupted when the Office of Pipeline Safety's personnel and resources are diverted to investigate a major pipeline failure.

The bill would also increase funding to States for the promotion of one-call program and expand the DOT role in pipeline safety to include pipeline safety awareness programs. Let the public know what they have to be concerned about and include them in the army of supervisors who will report any activity that they see to make sure that this—that the proper procedures have been followed and the call has been made to the one-call station.

The legislation directs the Secretary to establish an electronic data system on the nature, extent and geographic location of pipeline facilities in order to facilitate risk assessment and safety planning with respect to such facilities. The bill also increases inspection and siting requirements for pipelines in high density population areas, encourages residents in the vicinity of a pipeline to report suspicious dumping or, as I said earlier, digging on a pipeline right-of-way and make illegal dumping on a pipeline right-of-way a Federal crime.

Mr. Chairman, I believe that this bill and the comprehensive one-call bill provide positive steps toward addressing the very real issue of pipeline safety in an era of aging pipeline infrastructure and increasing demand. We cannot and should not plan to do away with this very important energy source. What we have to do is make certain that as we employ this valuable resource, that we are not endangering residents or businesses or transportation facilities near these pipelines, and I think that we can do it. We have lots of transmission pipelines and then we have the distribution lines, which I am sure you are aware of those, that feed specific areas.

In the Edison explosion, it was the transmission pipeline that blew up. The other services were commenced within a couple days after the tragedy took place. With that, Mr. Chairman, I say thank you very much for permitting me to testify before you and I hope that we will get quick action on this bill.

Mr. SHARP. Well, thank you very much, Senator. We certainly share your concern and that of the New Jersey delegation. You and others have worked vigorously and rapidly to try to address what issues we can to stem any further problems for other citizens around the country and we appreciate that.

Let me recognize the Ranking Member of the full committee, Mr. Moorhead, if he has any questions.

Mr. MOORHEAD. Well, the only question I have, of course, is that California has very comprehensive legislation in this particular. It seems to be working very well. I wanted to be sure that the Federal law would not interfere with the well-working program that they have in our State, and I would like to know what your comments would be on that.

Senator LAUTENBERG. Well, as the distinguished representative knows, that this is a question that is often in debate: when and where does the Federal Government take over a State's functioning, well-functioning responsibility. We have that often on environmental questions in particular generally. I assure you that if it is not already inscribed in the bill, that we would make certain that in no way do we reduce the compliance rate or the effectiveness of existing services within the State, and my colleague, Congressman Pallone, came in and I would assume that since he has heard the question, that he would be in agreement that we are not going to preempt State functions that are doing the job.

Mr. MOORHEAD. I had heard the statement earlier that the purpose is to get States to enact legislation that would adequately take care of this issue, and if that is the purpose, then, of course, a State like California has taken care of the issue. And they shouldn't be interfered with.

Senator LAUTENBERG. Well, particularly we want to encourage the one-call facility so that when the call is placed, that there is an appropriate response from this central agency that says this is where you cannot dig or that if you check the map, so-and-so, you will see the areas very clearly delineated.

Mr. MOORHEAD. That could be a very serious problem if you have to go to too much effort.

Senator LAUTENBERG. Yes.

Mr. MOORHEAD. When you are building a home or you are making an addition or there is construction progress.

Senator LAUTENBERG. It shouldn't be tough, and on the other bill, what we want to do is to try and have within the DOT nationally is a mapping of all pipelines across the country. Now, that sounds complicated. It really isn't because the companies all have this information available or they are in serious trouble if they don't, but we want to make certain that the public is aware of what their responsibilities are, of whether or not they can be guaranteed safety with these operating pipes we are talking about.

Having earlier shut off on the valves than we were able to see in Edison, which took a couple of hours to get done so the gas supply was continuing to flow, when in fact if we had some kind of an automatic response, it might have been able to be cut down in minutes and maybe that plume wouldn't have been quite as high or extensive as it was. We saw cars that the paint burned off them. When the residents tried to reach them, they burned their hands.

Mr. MOORHEAD. Do we have an estimate on cost? Who is going to pay for it?

Senator LAUTENBERG. We don't have a specific estimate. What I am told is we are talking about grants of relatively *de minimis* proportion, that we are not looking at anything that entails substantial expense not now covered by the government, and part of what we want to do is we have pipeline inspectors out there, is when there is a special investigation, to be able to either contract or employ extra people so that we can get that job done, making the pipeline companies just a little more aware of the fact that they might have some extra expense that they are going to have to pay for so we don't have to get involved in extra costs of maintaining the standard service.

Mr. MOORHEAD. Are we talking about the one-call bill or——

Senator LAUTENBERG. Talking about both of them. We could separate them, yes.

Mr. MOORHEAD. The one-call shouldn't be too expensive.

Senator LAUTENBERG. No.

Mr. MOORHEAD. Does the other one have some bit of cost?

Senator LAUTENBERG. I don't think so. We call for pipeline inspections. They are largely supervisory functions. We call for the pigging of the pipeline, which is an instrument that runs through, and very effective, would have to be—reporting the development of the database could be one of the larger costs.

I spoke to the Secretary of Transportation. He was with us in New Jersey the other day, and it looks like this is relatively available—readily available software and we are not talking about a major expense.

Mr. MOORHEAD. We heard an estimate of $20 million from the department.

Senator LAUTENBERG. Over a period of time.

Mr. MOORHEAD. How much time?

Senator LAUTENBERG. I am sorry. I don't have the specifics, Mr. Moorhead, but I am sure when you told—as you told your hearings, that information will be relayed to you and we will be very interested in that testimony as well.

Mr. SHARP. Thank you very much. Mr. Pallone.

Mr. PALLONE. Well, I just wanted to say that Senator Lautenberg, of course, has been outstanding in taking the leadership on this issue and I never saw such an interest personally. He came there within a day, I think it was the next day or two of the accident and he has been—he has been helping residents and really showing a personal involvement in this, as well as sponsoring the legislation, so I just wanted to commend you again.

Thanks.

Senator LAUTENBERG. Thank you.

Mr. SHARP. Well, Senator, thank you very much. We will look forward to working with you in the weeks ahead.

Senator LAUTENBERG. Thanks very much.

Mr. SHARP. We will now call the third panel of witnesses. We have with us Mr. Richard D. Huriaux, director of the Office of Engineering with the DC. Public Service Commission. He is representing the National Association of Pipeline Safety Representatives. Mr. Phil Becker, president of Utilities Consolidated, representing the Association of General Contractors of America; and Mr. Paul DeVaney of Bellcore with the Bell Communications Research, and he is representing the Network Reliability Council.

Mr. Huriaux, we would be happy to hear from you now. As I think you are aware, we will put all your written testimony in the record and we will be glad to hear an oral summary at this point.

STATEMENTS OF RICHARD D. HURIAUX, ON BEHALF OF THE NATIONAL ASSOCIATION OF PIPELINE SAFETY REPRESENTATIVES; PHIL BECKER, ON BEHALF OF THE ASSOCIATION OF GENERAL CONTRACTORS OF AMERICA; AND PAUL DEVANEY, ON BEHALF OF THE NETWORK RELIABILITY COUNCIL

Mr. HURIAUX. Thank you very much, Mr. Chairman.

As I know you are aware but some of the people in the room may not be aware, the National Association of Pipeline Safety Representatives is the association of the chief pipeline safety officers of each of the States, District of Columbia and Puerto Rico, and we work closely as an agent of the Office of Pipeline Safety, as well as enforcing our own State and local regulations, which in some cases are more stringent than those at the national level.

It is wonderful to be in a hearing where there is really so much agreement, referring now to the one-call bill, on what the policy should be. Over the years, the pipeline industry and its regulators have made tremendous strides in pipeline safety and this is shown by the statistics, declining pipeline accidents due to dig-ups, third party damage, if you will, over the years.

Our view, NAPSR's view, is that this bill will take us the last 20 percent of the way home on this issue. The States and the Federal Office of Pipeline Safety have, we believe, taken us the first 80 percent.

I will not read my testimony. Rather, I would like to emphasize certain points that I think need to be recognized in the bill to make it effective and also to make it more easily enforceable.

First of all, many of the States have very effective one-call systems and have for many years. Nothing in this bill should in any way disrupt the operations of those systems. An example would be the discussions that have gone on over the last few months concerning a national one-call number which NAPSR does support. However, I believe that this number should be a clearinghouse number rather than the number being advertised locally for one-call operations.

This is because the various one-call centers in States have invested tremendous amounts of time and resources and advertising effort in promoting Julie, Blue Stake in Arizona, of course Miss Utility here in the Washington area, and we believe there is a fairly high public recognition in the local areas of these names.

National one-call number, however, would be very useful for contractors working outside of their normal areas. It would give them a place to call if there was any uncertainty as to which one-call center they should be contacting and so, I believe, could increase, give us an increment of safety in that area.

There has been some discussion of the Federal, State issues. I really believe that these aren't too severe. The States, through the NAPSR representatives, have been working closely with the Office of Pipeline Safety, local distribution companies, as well as interstate pipeline companies in promoting one-call systems and pipeline safety for many years. This is an established relationship and as you well know, Mr. Sharp, we already have 49 CRF 198 issued by the Office of Pipeline Safety of DOT that does provide minimum standards for one-call systems for subject facilities, pipeline facilities.

Now, the one area that this bill really addresses well, I think, is the matter of exceptions or exemptions from membership in one-call systems. Many of the States, State laws specifically exempt municipalities or State departments of transportation and other owners of perhaps half or more of the underground facilities and especially in urban areas, water, sanitary sewer, storm sewers, private communication systems. The removal of these exemptions, which as I read the bill would be one of the effects of it, have a salutary effect on compliance with even the existing one-call operations in the States.

And I don't believe that most of the one-call systems in the country would have great difficulty absorbing these additional underground facility owner members. In fact, I think generally it would also be welcomed by excavators in the States. After all, it is excavators who are in immediate danger when they hit a waterline or a gas line or perhaps even worse, underground electrical facilities.

Education is another area which the bill does address and I think there is some opportunity for the national government, especially through the Office of Pipeline Safety, to increase public information activities related to the promotion of one-call numbers in the various States.

However, I think each State should—varied as they are, should be left to set up their one-call systems as they see fit. Rhode Island is in a very different situation than a State such as Texas. One one-call center? Rhode Island obviously, it is the only obvious way to go, as it would be in my jurisdiction, the District of Columbia. However, in Texas, in California, in your larger States with widely separated urban areas, the necessity for a State-wide one-call number is really not clear, and I think, as I said before, the States and the Office of Pipeline Safety and the pipeline companies have brought the one-call issue about 80 percent of the way to full implementation.

So really what I think we should be focusing on is fine-tuning these efforts, and as members of the committee and witnesses have said, we shouldn't interfere unnecessarily with what on the whole has been a very successful program, but rather should focus on making it more successful.

In regard to the second bill, the National Association of Pipeline Safety representatives' members really haven't had a chance to express their opinions on that, but I am sure we would like the opportunity to do so in coming days as we proceed through markup and various amendments. I am sure we will be in contact and will have opinions on that bill, but obviously with 49 State level members we, in the last several days, haven't had time to poll them all, so I hesitant to speak on their behalf on that bill.

So in general, as our testimony indicates, NAPSR is in support of this bill, the one-call bill, H.R. 4394, and we stand ready to work with the committee, as we have in the past, in making whatever final adjustments may be necessary to address the State's concerns.

Thank you very much.

Mr. SHARP. Thank you very much, Mr. Huriaux. I am very pleased now to recognize Senator Bradley who can be with us at this point. He has long been interested in gas pipeline safety issues and held a hearing of his own in New Jersey following the accident and we appreciate his input at this point.

STATEMENT OF HON. BILL BRADLEY, A U.S. SENATOR FROM THE STATE OF NEW JERSEY

Senator BRADLEY. Thank you very much, Mr. Chairman. I appreciate the opportunity to be here and to have a chance to testify before you, distinguished members on the Republican side and my distinguished colleague from New Jersey, Mr. Pallone, who spent as much time as—well, not quite as much time at the explosion site as the residents, but he was there a lot, and it is kind of an indication of his own dedication to the welfare of his constituents, so it is a pleasure to be here and to testify on behalf of the comprehensive one-call legislation.

The bill you are considering will create new assurance that accidents involving pipelines and underground utilities won't occur. Every year multiple facilities and tens of millions of dollars worth of damage occur simply because people dig where they shouldn't dig. These third party incidents are the single leading cause of accidents involving pipelines.

According to the Department of Transportation, these accidents result in over half of the fatalities and half of the property damage caused by all pipeline failures. Legislation that I introduced in the Senate, the Comprehensive One-Call Notification Act, will create a mechanism to prevent the inadvertent injury and the potential tragedy.

Last March 23rd, just before midnight, an explosion ripped through the community of Durham Woods in Edison, New Jersey. Within minutes, eight apartment buildings were ablaze. Within minutes. Soon they were gone, wiped out by a fireball that lit up the sky over 100 square mile area. One life was lost. That is all. Hundreds lost their homes. Many more were evacuated.

The injuries, as I said, were miraculously low, but who knows how many others still lie awake at night wondering whether it could happen again and fearing the future? I have personally talked to a number of those residents who have talked about their being trapped, thinking they are being trapped, going against the door, breaking through the window just to get out and who have recurring nightmares because of that.

Reflecting on the accident today, it seems hard to fault anyone for their response to the tragedy. This was one of those occasions where pointing the finger just wouldn't work. The community pulled together to help out those in need, food, emergency shelter, general support, financial assistance were offered amply and unconditionally in the hours and days following the accident. However, great as this response was, this was not what, from my perspective, was the most striking aspect of the accident.

What is most striking is how lucky we were. Who would ever think that given the timing and the magnitude of the explosion, so many people—many fleeing, as I said, with just their clothes on—would escape without serious injury.

Few who have walked around that crater, who have seen the melted cars, the empty building foundations, the melted shingles on the roof of the apartment buildings that still remain, would disagree with the conclusion that many there were saved only by a miracle. Unfortunately, miracles are a poor basis for public policy. I haven't yet discovered how to summon them when I need them and as any elected official knows, we need them on a regular basis. At least every 2 or 6 years, because you can't count on them.

I am not about to count on them and the fact is that there is no margin for error in today's pipeline industry. The natural gas industry does have an excellent safety record, especially when you consider that 25 percent of the energy we now consume moves through these pipelines. For example, there are several major pipelines that cross New Jersey and hundreds of smaller ones, but the Edison accident never should have happened.

We need to acknowledge Edison for what it was, a breakdown in the regulatory and safety program. When the National Transportation and Safety Board testified before the Energy Committee in April, their analysis pointed nearly conclusively to multiple gouges on the pipeline as the possible cause of the disaster. You can actually see the damaged pipeline. These marks appeared to be due to some powerful machinery, such as a backhoe that dug into the ground and hit the pipeline there by several millimeters, reducing the thickness of the pipeline.

At this point, we don't know whether the damage was unintentional or on purpose. We don't know who struck the pipeline or whether they might have been aware of the possibility. We do know, however, that there was no requirement of utility notification prior to the excavation and we know that there is no penalty for digging in the vicinity of a pipeline without notifying the utility operator.

This is simply wrong and represents a failure of public policy.

At the hearing I held in April, every witness, and I mean every witness, agreed that we need a new national program of utility notification. If someone is excavating or grading a site, there has to be proper notification and it has to be mandatory, not voluntary, without exceptions and with penalties for negligence or noncompliance.

This national program will be created by comprehensive legislation that you are considering today. Right now the gas industry is making plans for rapid expansion into new markets, particularly in the areas of natural gas vehicles and electric power production. The Department of Energy has predicted that the gas market will expand by ⅓ over the next 15 years.

If accidents occur, regardless of who is at fault or how the industry follows up, this growth will not happen. It is as simple as that.

In one sense, this bill is unnecessary. Sooner or later I predict every State will have a one-call provision like those identified in this legislation. The reason is simple. Sooner or later every State will experience a major accident involving third party damage to underground facilities, then, just as happened in New Jersey, one-call provisions will be introduced or strengthened.

This is not an issue of cost. Most States have these programs, as I said, already. The problem is that absent sufficient political motivation, the programs are just not as effective as they need to be. We shouldn't have to wait for another disaster to understand the importance of this modest bill.

This comprehensive one-call legislation represents a necessary step if we are to do everything reasonable and appropriate to prevent the public from the kind of tragedy that hit Edison. The bill obviously won't guarantee that another Edison will never occur, but mandatory, loophole-free, one-call programs will send a message to the public that our concern is serious and our risks are real. This awareness would be a powerful ally in our fight for increased safety.

Mr. Chairman, I say to you and to the committee, I encourage you in the work to pass this legislation. I know that many of you are very supportive and I appreciate that and I believe that there is no substantive reason why we can't and should not act, and I am prepared to work in the Senate with the Commerce Committee to gain approval of the legislation. I intend to do that. I hope we can work together to make this concept a public law in this Congress, a public law that will give people some greater certainty that the fireball that exploded in Edison last March won't happen in their neighborhood.

Mr. SHARP. Thank you very much, Senator. We appreciate your serious concern about this and your work on the issue and look forward to working with you in the weeks ahead.

One of the issues that we are grappling with is the technique by which we deal with the State governments in getting them to install or improve their programs, and so we are welcoming from everybody the technique that seems to get the job done and yet is the least burdensome or at least most acceptable to the State governments.

We obviously want to avoid the issue of unfunded mandates which most of us do not think should arise in this bill, but is one which the governors and others are very, very sensitive to. Anytime we impose a cost on their taxpayers that they have to pass on, this concern is raised even though as I understand it, that should not be necessary whatsoever. We want to make sure we are up front on that issue before ever getting started so that we can be successful in getting these programs in place.

I just raise that with you so your folks as they have ideas will welcome it. The more we can coordinate across the chambers, the faster we can get things done.

Senator BRADLEY. Be great to have a bill that didn't need a conference.

Mr. SHARP. That is right, exactly. You and others might have need to be in a health care conference, I hope.

The gentleman from California, the Ranking Member of our full committee, Mr. Moorhead.

Mr. MOORHEAD. Senator, we appreciate your work on this issue and will be listening to the testimony of the other witnesses here today, but we want to thank you.

Senator BRADLEY. Thank you, Mr. Moorhead. Thank you for your leadership.

Mr. SHARP. Mr. Pallone, do you have any——

Mr. PALLONE. No. I just wanted to commend the Senator once again, and I know you are determined to have the bill passed and to work with the House to try to get it passed and that is really great. I think that is the most important thing in response to the accident.

So thank you.

Senator BRADLEY. I think the Sharp-Pallone bill must pass—I am sorry, Sharp-Pallone-Moorhead bill.

Mr. PALLONE. Thanks.

Mr. SHARP. I see why the Senator is so successful.

Thank you very much, Senator, we appreciate your coming over.

Mr. SHARP. We are under the call of a potential other vote. My understanding is we have 15 minutes on the quorum call prior to the 5 minute vote, so what I think we will try do is take Mr. Becker and see if we can complete his oral testimony before we have to rush out of here.

Mr. Becker, we would be happy to recognize you at this point.

STATEMENT OF PHIL BECKER

Mr. BECKER. Thank you, Mr. Chairman, members of the committee. My name is Phil Becker and I am an underground utility contractor from San Antonio, Texas. I am here today representing the Associated General Contractors of America, AGC. With me in the audience is Brian Deere, member of the national staff.

AGC is a national association of 30,000 construction industry businesses including 8,000 of the Nation's leading general contracting firms. AGC has 100-plus chapters nationwide which represent the interests of the construction industry on the local and State levels. AGC members are involved in all types of construction activities.

On the national level, AGC has had experience with the creation and implementation of a model underground facility protection program. In 1989, AGC worked with the Electronics Industries Association and others on a standard for the protection of underground fiber optic cable.

Because a vast majority of construction projects involve some type of excavation, AGC strongly supports universal participation in State one-call systems as called for in H.R. 4394. With the millions of miles of pipelines, water and sewerage pipes, telephone, electric, cable TV and fiber optic networks hidden underground, it is important that contractors have an understanding of what lies beneath the surface.

This knowledge is important to protect the contractor from liability, to protect the employees from unknown hazards, and to protect the general public from physical harm and disruptions to their utility service. Having as much information about the nature and location of these underground conditions is extremely important to the contractor.

The primary concern is, of course, safety. Contractors' employees can face significant dangers when they are undertaking excavation activities, particularly if underground utilities have not been properly located. These dangers include explosion, fire and electrocution. AGC and AGC members consider the safety of employees a number one priority in the industry.

AGC has developed dozens of videos, training manuals, and other materials to help contractors provide a safe workplace and to teach employees how to work safely.

AGC's comments on specific proposals contained in H.R. 4394 follows. The national one-call telephone number. We commend representative Pallone for proposing a national one-call number in the legislation. We are hopeful that it will be possible to obtain a simple, easily remembered number, such as a 411 or 911 number currently utilized for telephone directory information and police and fire emergency. It is AGC's feeling that if a simple user friendly system is established, it will increase voluntary participation in the program.

If such a number results from an enactment of this legislation, it is our hope that calling the one-call number will become a natural reflection for contractors prior to commencing excavation. It is our understanding, however, that the proposed national number will serve merely as a conduit for directing calls to the appropriate one-call system in the State where the excavation will occur. For this concept to work, the response time of the State one-call system must be at least as quick as it is currently.

State one-call legislation. AGC supports the concept of requiring States to enact one-call legislation rather than impose a Federal standard. It has been AGC's experience, based on participating in the process of formulating one-call bills in numerous States, that each State should have the opportunity to enact legislation that is sensitive to the industries, practices and experiences in that particular State. Many States have excellent one-call systems and AGC is hopeful that the bill will not undermine any of these existing programs.

However, AGC is opposed to this suggested sanction that be applied to States that do not establish a system within 3 years. The draft bill would withhold ½ of 1 percent of a State's Federal highway allocation for failure to enact a one-call system. Not only would this penalty deprive a State of much needed assistance for transportation needs, it also deprives the motoring public who pay user fees in the form of gasoline taxes with a return on their investment, and it deprives highway construction workers of employment opportunities. In addition, a withholding of Federal highway funds would adversely impact highway and bridge safety improvement programs.

Finally, the proposed sanction may force many States to weigh the costs of ½ of 1 percent reduction in Federal highway funds against the cost anticipated by local governments in participating in the one-call system. That choice would result in either a lost opportunity to improve underground facility safety or less investment in highway and bridge improvements designed to enhance safety. AGC prefers as an alternative some type of incentive program, such as grant funds to aid in setting up the system, be used instead.

Timeliness of calls. AGC believes that 48 hours is the maximum amount of time that should be required for notification to be given by the excavator and for facility owners to respond before excavation activities begin. A 48-hour waiting period is generally what is required under existing State and local programs. Construction activities cannot be so precisely timed that a contractor can always project more than 48 hours out when an excavation will occur in a specific area. A variety of factors can affect the construction schedule.

It should be noted that it is often necessary to locate utilities throughout a construction project, not just at the start of it. AGC is hopeful that facility owners who may be affected by the excavation respond in a timely fashion and not wait until the last minute and beyond.

Time is a critical element in the construction process. Construction delays can significantly impact the progress of the work and ultimately the success of the project. For this reason, contractors must be concerned about the timeliness of locating underground facilities before construction begins and about potential delays if an unanticipated underground condition is found once construction is underway. A response within 24 hours would be more ideal.

Mandatory participation. AGC supports efforts to require all underground facility operators, including State and local governments, to participate in a one-call system. For a one-call system to be truly effective, it is vital that all underground facility owners participate. While AGC is sensitive to arguments that we anticipate will be advanced by local government entities that they would find it financially difficult to comply with requirements of the legislation, AGC believes that if there is to be any confidence in the system, all owners of underground facilities should be required to comply with the law. Although some may argue that this requirement is an unfunded Federal mandate, AGC believes that it is the responsibility of all underground facility owners to protect their facilities.

The penalties. AGC believes that one-call systems can only work if they operate in the spirit of cooperation and fair allocation. To be successful, all participants must fulfill their responsibilities. Contractors must call on timely fashion, wait the appropriate amount of time for a response and then faithfully follow the markings. Owners must respond to the call in a timely fashion and the markings must be accurate.

AGC is hopeful that this new effort will build on the successes of existing programs through the use of educational efforts and outreach rather than through a heavy handed sanction approach.

AGC is also hopeful that the program will not become a revenue-raising exercise. AGC is concerned about the potentially excessive size of the penalties contained in the proposal.

As we read it, a single violation could result in a $25,000 fine. AGC suggests that the penalties be based on willful violations and that they be applied only on a graduated basis. By this we mean that penalties should increase only where there is a pattern of willful violations.

AGC also suggests that a good faith standard be included to mitigate situations when a party has made a good faith effort to comply with the requirements of this legislation.

Mr. SHARP. Mr. Becker, I am sorry we are going to have to interrupt you. We are down to about 4 minutes to vote. If I can get to you just in 30 seconds finish up your testimony, if—if there is some quick point you want to make in the last 30 seconds, I will be glad to hear it and then I will have to bring the committee to a break here.

Mr. BECKER. Yes, sir, in conclusion, we believe that the one-call concept pledges to work—that we do pledge our association to work with the Congress in fashioning a bill that is fair to everybody, to all parties, that is user friendly, and I would like to emphasize that training and education is what it is going to take to make this bill work and not heavy-handed penalties.

Mr. SHARP. Thank you very much, Mr. Becker. We are going to take a break and we will return. We have two votes in a row so it will probably be 10 to 15 minutes by the time we are back here and then we will take Mr. DeVaney at that point. Thank you.

[Breif recess.]

Mr. SHARP. The subcommittee will please come to order. We potentially face a series of votes on the House Floor, and so we are going to try to expeditiously complete the panel. Mr. DeVaney, we will be happy to hear from you at this point

STATEMENT OF PAUL E. DeVANEY

Mr. DEVANEY. Thank you, Mr. Chairman and members of the subcommittee.

As a member of Bellcore technical staff in New Jersey, Bell Communications Research, it is my responsibility to provide technical standards contributions, consultation and interpretations related to telecommunications facilities. I actively participate on a national level with associations, Federal agencies, and standards bodies involved with rights of way, damage prevention, legislation and regulatory matters, as well as other industry related activities that have impact on telecommunications network operations.

I have previously served as a member of the Network Reliability Council, Fiber Cable Focus Team, and it is in this capacity that this statement is being provided today. Let me thank you for the opportunity to provide this statement from the Fiber Cable Focus Group. Unfortunately, the chairman of the group, Mr. Daniel E. Crawford, the president of network services of MCI Communications Corporation was unable to be presented today. It is my understanding that he has supplied a written statement for the record and has provided as an attachment to his article, "Fiber Optic Cable Dig-Ups, Causes and Cures", published in 1991. I brought with me a sample of a fiber optics cable that we have passed out to the subcommittee.

Today's fiber network consists of more than 325 sheath miles of cables containing approximately 10 million miles of optical fibers. While the total telecommunications network consists of well over 3.5 million sheath miles of both fiber and copper cable, the significant advances in fiber technology have resulted in increasing concentrations of data being funneled through the fiber cables.

The significant consequences of a single cut have increased with reliance on new communications technologies which are revolutionizing the telecommunications industry. Voice and data transmission over fiber cables has become a essential means of communications for a wide range of organizations and interests, including hospitals, airports, police departments, as well as financial institutions and business in general.

Fiber optic cables, whose size is often less than an inch in diameter, routinely carry tens of thousands of telephone calls over glass strands each the size of a human hair. Therefore, protecting these vital links takes on ever-increasing importance.

After several major communications outages during the second half of 1991, the Federal Communications Commission established an advisory committee to enhance network reliability. The advisory committee, called the Network Reliability Council, NRC, brought together leaders of the telecommunications industry and telecommunications experts from academic and consumer organizations to explore and recommend procedures that would enhance network reliability.

The council established focus teams comprised of experts both inside and outside the telecommunications industry to conduct in depth studies of seven network reliability areas. One of the areas that were considered to be of high priority based on historical data was fiber optic cable systems.

Today's testimony will focus on the results and recommendations made by the fiber cable system group as they pertain to the one-call legislation.

The mission of the group was to look at the causes of fiber cable outages and the potential opportunities to minimize these incidents. Utility company records, industry surveys and FCC data consistently reveal that cable dig-ups account for a large portion of the cable failures and are caused largely by construction site excavation activities.

The group collected data pertaining to service disruptions, a definition of a failure occurrence is; an event that caused the system to fail to perform to specification and requirements, or an event that necessitates repair action or component replacement in the immediate or near future. The Focus Group data collection period covered all FCC reportable field failures—a reportable field failure is 50,000 customers out of service for more than 30 minutes, between April 1st, 1992 and August 31, 1992 and all fiber cable failures that occurred between September 1st, 1992 and February 12, 1993.

Understanding the causes of fiber optic cable failure is essential for identifying effective damage prevention methods. Understanding the relative significance of these causes and the influences on cable failure probability are necessary for targeting damage prevention initiatives and for formulating a complete, comprehensive facility protection strategy. Various analytical methods were used to arrive at the Focus Group's conclusions. The data analysis was conducted on the field failures reported and augmented with data from a field tracking study.

The results of this analysis on all fiber failures concluded that dig-ups are the single largest cause of fiber optic cable damage accounting for more than 58 percent of the reported failures. The next largest cause represented 7 percent. With the findings that dig-ups are the largest cause of fiber optic cable damage, therefore their mitigation embodies the most significant opportunity for cable reliability improvement.

Utilizing the reported data, a detailed root cause analysis was performed and a probable root cause assigned. The leading root cause, 39 percent of reported cable dig-ups is excavator error made during digging. These failures occurred after the facility owner was notified and the route properly located and marked.

The second largest cause, 33 percent, is the excavator's failure to provide notification prior to digging.

As the results have indicated, a substantial portion of dig-ups occur because an excavator fails to notify the facility owner prior to digging. A larger number of dig-ups occur despite notification by the excavator, accurate cable locating and the presence of cable markings.

The Network Reliability Council, Fiber Cable Focus Group, provided extensive details pertaining to causes, activities and recommended solutions that could reduce the persistent industry-wide problem of dig-ups, as well as other fiber cable damage. The information was presented at the Federal Communications Commission's network reliability report to the Nation in June 1993.

Based on the findings and discussions, the group recommended:

Number one, the strengthening of utility damage prevention legislation, call-before-you-dig enhancement.

Number two, broad and uniform implementation of "best practices" to minimize cable damage, standardized practices and procedures.

Endorsement of a benchmarking study to identify innovative approaches to fiber cable damage prevention. Best practices provide a certain level of assurance in cable damage prevention, yet cable damage continues to occur. A benchmarking study is now underway to search for these innovative approaches to damage prevention. The study's objective is to identify the most successful policies, strategies and implementations for installing, maintaining and protecting fiber optic cable from major causes of damage.

The telecommunications industry uses a national standard known as the ANSI standard EIA, TIA-590 to bury and put their cables in the ground and uses that for their marking. So there is a national standard for the basis to do fiber optic cables.

As previously stated, dig-ups are the largest cause of fiber cable failures and account for 60 percent of the failures reported. Analysis indicated that 33 percent of the reported dig-ups resulted from the excavator's failure to notify the facility before digging started. Although over 40 States have damage prevention laws requiring such notification, the laws are generally weak, not adequately enforced and provide for punitive damages in the event of an excavator negligence. Therefore, the group recommends enforcement, enactment and/or revision of Federal, State and local damage prevention laws.

A Network Reliability Steering Committee, under the auspices of the Alliance of Telecommunications Industry Solutions, was formed after the NRC gave their report to monitor network reliability utilizing major outage reports as per FCC docket 91273. The committee's charge is to analyze the industry's reports of network outages to identify trends, distribute the results of its findings to the industry, and where applicable, refer matters to appropriate industry forums for further resolution, to help ensure a continued high level of network reliability.

The report for the fourth quarter of 1993 stated "all failure categories, local switch, tandem, facility, power, overload, are within the control limits." While facilities was again the dominant category, the actual number of outages was much lower than that experienced in the third quarter, cable dig-ups continue to be the single largest cause of facility damage with seven of these occurring because the affected carrier received no notification prior to digging. So we support and have since the June 1993 report to the Nation damage prevention and one-call or call-before-you-dig legislation.

[Testimony resumes on p. 179.]

[The prepared statement and attachment of Mr. Daniel Crawford follow:]

WRITTEN STATEMENT OF
DANIEL E. CRAWFORD
PRESIDENT, NETWORK SERVICES
MCI COMMUNICATIONS CORPORATION
BEFORE THE
COMMITTEE ON ENERGY AND COMMERCE
SUBCOMMITTEE ON ENERGY AND POWER
UNITED STATES HOUSE OF REPRESENTATIVES
JUNE 23, 1994

Mr. Chairman and Members of the Subcommittee:

I am Daniel E. Crawford, the President of Network Services of MCI Communications Corporation (MCI). In that capacity, I oversee all technical and operational aspects of MCI's network, including repair of severed or otherwise damaged cable. I have previously served as the head of the Network Reliability Council's (NRC) Focus Team on Fiber Cable Systems and was the principal author of the article "Fiber Optic Cable Dig-Ups: Causes and Cures" which was published in the NRC's June 1993 report "Network Reliability: A Report to the Nation." I am submitting that article as an attachment to my testimony and ask that it be made part of the hearing record.

On behalf of MCI, I wish to thank the Subcommittee for affording me the opportunity to submit this statement regarding H.R. 4394, the "Comprehensive One-Call Notification Act of 1994." MCI commends you, Mr. Chairman, Representative Pallone and the entire Subcommittee for your efforts to fashion responsible legislation to address the concerns stemming from dig-ups of underground facilities, including communications facilities.

MCI celebrated its 25th anniversary in 1993. During those years we converted a network which began with a single microwave link between Chicago and St. Louis to one which now provides service to the nation over high capacity optic fiber. MCI's network now utilizes 21,000 miles of optic fiber cable nationwide, and approximately 97% of that fiber is in underground facilities. In the vigorously competitive long distance market, MCI must strive for "bulletproof" network reliability. We simply will not survive in a competitive environment if our network is down.

We view H.R. 4394 as a particularly important initiative because we believe that although our fiber network is underground and out of sight it should never be out of mind. The optic fiber located in these underground storage facilities, although composed of glass strands only slightly thicker than human hair, nonetheless comprises the backbone of the much-heralded information superhighway. Clearly, as our economy becomes increasingly information based, maintaining the integrity of the highways upon which that information flows takes on an added degree of importance in furthering our nation's economic progress. Outages related to cable failures cost carriers and customers alike and take time to repair. The industry average time for service restoration is 5.2 hours and the average time for cable repair is 14.2 hours.

More importantly, and just as with pipelines and other utilities which occupy underground facilities and rights of way, ensuring the integrity of underground communications facilities has broad-based public heath and safety implications. Whether related to encrypted communications of the United States Armed Forces, telemedicine, nuclear power facilities or other vital societal functions, an interruption in service precipitated by a fiber cut can dramatically affect the lives of millions of Americans.

The Fiber Cable Focus Group of the NRC, which I headed, collected and analyzed data on service interruptions traced to cable failures caused by dig-ups. The Group also reviewed the practices of other utility companies and outside groups such as "one-call" associations. The first recommendation made by the Group was to strengthen utility damage prevention legislation and enforcement -- so called "call-before-you-dig" or "one-call" legislation.

The priority given call-before-you-dig measures can easily be accounted for. Nearly 100% of network failures result from a broken or damaged cable. Dig-ups are the largest single cause of fiber cable failures and account for nearly 60% of the cable failures reported in the industry. When we examined the root causes of the dig-up occasioned fiber cuts, we noted that 33% resulted from the excavators' failure to notify the owner of the underground facility before digging started.

MCI's internal numbers indicate that properly administered call-before-you-dig systems are both effective and efficient in drastically reducing the number of outages resulting from cable cuts. In 1993, for example, MCI received approximately 569,365 notifications from call-before-you-dig or one-call systems. We sent engineers to the field tc locate cable on 79,711 occasions. Of that number of excavations, only once was MCI's cable damaged.

However, federal legislation is needed to ensure that call-before-you-dig systems are uniformly developed and administered across the country. Although over 40 states in 1993 had damage prevention laws requiring notification prior to excavation, the Fiber Cable Focus Group found these laws were generally weak, not adequately enforced and provided little in the way of deterrence through punitive damages. That is why federal legislation is so necessary.

Let me demonstrate the pressing need to address the one-call issue on the federal level through some examples. A new one-call statute in Kentucky exempts all utilities and their contractors from one call compliance. This amounts to approximately one-third of all excavations statewide. In Georgia, North Carolina and South Carolina, the states' Department of Transportation are exempt.

We view statutes with these kinds of broad exemptions as creating a situation worse than having no statute at all. Al least, in the absence of a statute, there is an industry standard to call before digging. But granting exemptions only promotes more reckless excavations since anyone harmed as a result of a dig-up otherwise in violation of the state one-call statute has no recourse.

MCI believes that all excavators should be equally subject to call-before-you-dig requirements. The Fiber Cuts Focus Group found that excavators working for local telephone companies, electric, gas, water and highway department contractors were all involved in significant degrees with cable cuts. Only approximately 4.5% of cable cuts resulted from dig-ups by private landowners.

H.R.4394 has played an important role in alerting industry to the importance of one-call issues, and the bill strikes a number of important balances. It seeks standardization of one-call systems across the country, not by establishing another federal bureaucracy, but rather, through setting minimum standards for state one-call systems. Each state, however, remains free to implement its own system once approved by the Secretary of Transportation. H.R.4394 imposes obligations on both the excavator and the facilities operators and in this regard, the bill establishes a framework for their cooperation.

However, in the absence of voluntary cooperation, H.R. 4394 also provides for proper penalties for repeated violation of the one-call provisions. This form of deterrence is vital to the success of any program designed to require excavators to call before they dig.

MCI is enthusiastic regarding the prospects for call-before-you-dig legislation and looks forward to working with you, Mr. Chairman, Mr. Pallone and each of the Members of the Subcommittee to further refine H.R.4394. Toward that end, we would be happy to supply the Subcommittee with whatever additional information or resources it may require as it continues to consider this important legislation.

Thank you.

FIBER OPTIC CABLE DIG-UPS - CAUSES AND CURES

Mr. Dan Crawford
President, Network Services
MCI Telecommunications Corporation
1801 Pennsylvania Avenue
Washington, D.C. 20006

1. Executive Summary

1.1 Overview and key messages.

As sub-committee member on NOREST, the Fiber Cable Focus Group collected and analyzed industry data concerning service interruptions resulting from cable failures caused by dig-ups. The Group also reviewed the practices of other utility companies and outside groups such as "one-call" associations. Through these activities and a statistical analysis of industry data, the Group has determined that several primary activities can mitigate or certainly reduce cable dig-ups.

Fiber optic cables, whose size is often less than one inch in diameter, routinely carry tens of thousands of telephone calls over glass strands slightly more thick than a human hair. Damage to fiber optic cables can shut down vital communications links to airports, emergency services, and nuclear power facilities. With advances in SONET transmissions and ATM switching technologies comes higher and higher concentrations of traffic placed on fiber cables. Therefore, protecting these vital "information highways" takes on ever increasing significance. To improve the reliability of these critical high capacity links, the Network Reliability Council's Fiber Cable Focus Group collected and analyzed fiber cable failure data, surveyed existing utility damage prevention legislation and held discussions with key representatives of the damage prevention industry.

Based on these findings and discussions, the Group recommends: 1) a strengthening of utility damage prevention legislation - call-before-you-dig enhancement, 2) broad and uniform implementation of "Best Practices" to minimize cable damage - standardized practices and procedures, and 3) endorsement of a benchmarking study - to identify innovative approaches to fiber cable damage prevention and/or assess the need to revise existing practices.

Damage Prevention Legislation

Dig-ups are the largest cause of fiber cable failures and account for nearly 60% of the failures reported by the industry. Examining the root causes of fiber cable dig-ups reveals that 33% of reported dig-ups resulted from the excavators' failure to notify the facility owner before digging started. Although over 40 states have damage prevention laws requiring such prior notification, the laws are generally weak, not adequately enforced, and provide little provision for punitive damages in the event of excavator negligence. Therefore, the Group recommends enforcement, enactment, and/or revision of federal, state, and local damage prevention laws. Paragraph 9.1 of this paper details the intent of this legislation.

Best Practices

Strong damage prevention legislation is essential for reducing the number of cable dig-ups. However, 40% of the reported dig-ups occurred in spite of prior notification by the excavator, accurate cable location, and proper temporary marking of the sub-surface cable route. Therefore, in addition to recommending strengthened damage prevention legislation, the Group has identified a compilation of best practices as procedures which, if uniformly complied with and broadly implemented, can be effective in minimizing occurrences of fiber cable dig-ups. These practices are grouped into four general areas:

1) engineering and construction,

2) call-before-you-dig,

3) effective cable location, and

4) additional preventive measures.

Although many of these procedures are currently specified in industry standards, they were generally not practiced in the failure events reported by the industry.

The Group found that although dig-ups were responsible for 70% of sub-surface cable failures, several other failure causes were consistently reported. While having the same probability of failure as sub-surface cable, aerial fiber optic cable was particularly susceptible to damage by a wider variety of causes. To prevent cable damage from causes other than dig-ups, the Group identified a broad range of practices which can be employed. These practices include additional personnel training to reduce accidental damage during cable maintenance or mining activities, installation techniques to minimize potential rodent damage, and other procedural and engineering practices.

Benchmarking Innovations

As the security of fiber optic cable becomes more critical to assuring telecommunications networks reliability, the Group recognizes the necessity of continuing to search for innovative approaches to damage prevention. To help accomplish this, the practices of other industries, in addition to those in telecommunications, must be continuously evaluated for their applicability to current preventive measures. Therefore, the Group recommends an industry-wide benchmarking activity to identify the most successful of these approaches and to highlight effective implementations of established and effective practices. The study would identify the most successful policies, strategies, and implementations for installing, maintaining, and protecting fiber cable from all major causes of damage.

The Group recommends that the benchmarking project be managed through the TIBC. These findings would be shared throughout the industry by the ECSA.

2. Background

2.1 The Fiber Cable Focus Group - motivation, goals and objectives.

Protecting fiber cable from physical damage is critical to ensuring the dependability and reliability of the public telecommunications network. During 1992, fiber cable failures were the single largest cause of network outages affecting more than 50,000 customers for more than 30 minutes and accounted for roughly as many outages as tandem and local switch equipment failures combined. As networks place ever increasing amounts of traffic over single fiber cables, the need to protect these vital facilities becomes ever more clear and crucial to the industry's goal of raising systems reliability.

2.1.1 Focus Group motivation.

The NOREST mission statement: "Look at the causes of fiber cable outages and the potential opportunities to minimize these incidents."

Notwithstanding the NOREST mission statement's critical objectives, recurring dig-up problems - validated by industry data - provide vigorous justification for and motivation to sustain Focus Group efforts in reducing cable dig-ups.

Utility company records, industry surveys, and FCC data consistently reveal that cable dig-ups account for more than 50% of all cable failures and are caused largely by construction site excavation activities. Therefore, in an effort to raise systems reliability through reduction of cable dig-ups, this paper will focus on renewed approaches to complete prevention, or certainly minimizing, the occurrence of this persistent industry-wide problem. Appendix 4 provides the NRC's Issue Statement concerning areas of study and preventive measures to improve the security and reliability of fiber optic cable.

2.1.2 Focus Group goals and objectives.

With ever increasing awareness of the necessity to protect fiber optic cable, NOREST established the Fiber Cable Focus Group to apply its expertise and ingenuity to define, develop, and implement preventive measures aimed at greatly reducing or ultimately eliminating cable dig-ups.

As envisioned, such measures and/or methodologies - whether new or improvements to existing ones - are the best, perhaps the only, means to increase systems reliability through the significant reduction of cable dig-ups. Preventive measures developed by the Focus Group were presented to NOREST as recommendations and incorporated a balance of legislative, technical, and procedural (e.g., call-before-you-dig) elements.

Focus Group sanctioned preventive measures will have the promise of reducing physical risks to fiber cable and can indirectly have other beneficial effects as well.

2.2 Organization of this presentation.

The basis upon which this paper is developed is the statistical study and methodology used to collect and analyze the Focus Group's request for industry data concerning service interruptions due to fiber cable failures. Study-based conclusions are presented for industry assessment as possible candidates for legislative and standards acceptance. The paper is organized as follows:

Section 1, **Executive Summary** - The Focus Group's connection to NOREST, study objectives and activities performed to develop this paper and its conclusions, and suggested actions to achieve increased network reliability are presented.

Section 2, **Background** - Explains the impetus for the Focus Group and its work, NOREST as the catalyst, and Group goals and objectives.

Section 3, **Team Membership** - Presents personnel and member organizations comprising the Fiber Cable Focus Group sub-committee to NOREST.

Section 4, **Data Collection and Analysis Methodology** - Introduces and describes the study's statistical methodology applied to identify and isolate factors/causes which consistently show they contribute to or directly result in cable dig-ups.

Section 5, **Causes of Fiber Optic Cable Damage** - The definitions of cable failure types and causes as used in this paper and definitions of root causes are presented, the definitions also mark the study's boundaries since they are the basis and the scope of cable failures covered by the Focus Group's efforts.

Section 6, **Key Lessons Learned and Recommended Best Practices** - Statistically-based conclusions, reached after industry data was collected and closely analyzed, are presented. Best practices are proposed along with other options ready for additional assessment by industry, standards, and legislative processes.

Section 7, **Metrics** - Explains that specific metrics will be offered as a result of a benchmarking effort.

Section 8, **Path Forward** - Demonstrates the case for continued monitoring by the ECSA, work still required to develop effective industry policies and procedures to achieve reductions in cable dig-ups is presented.

Section 9, **Conclusions** - Describes the need for assessment of current laws and procedures and suggests an industry-wide strategy aimed at reducing cable dig-ups as well as better serving utility industries at large.

Section 10 - **Acknowledgements**

Section 11 - **Reference**

Section 12 - **Figures and Exhibits**

Section 13 - **Appendices**

Glossary of Terms - Abbreviations and Acronyms

3. Team Membership

The Fiber Cable Focus Group team representatives include:

- Dan Crawford, MCI (Chairman)

- Frank Ianna, NOREST Champion

- Ron Binz, NASUCA

- Bob Crawford, Bell Atlantic

- Paul Devaney, Applied Concepts Inc. (formerly with Bellcore)

- Sam Lisle, Fujitsu (formerly with Bellcore)

- Mike Michalczyk, Southwestern Bell

- Peter Shelus, AT&T

4. Data Collection and Analysis Methodology

4.1 Data collection emphasis and goals.

The questionnaire employed by the Focus Group for collecting fiber cable damage data parallels the format and content of one developed by Bellcore. This questionnaire is shown in Appendix 1. It is utilized to collect data about service disruptions meeting certain criteria and definition requirements of a failure occurrence, that is: an event that caused the system to fail to perform to specifications and requirements or, an event which necessitates repair action or component replacement in the immediate or near future.

The Group's data gathering effort is patterned after an ongoing data collection and analysis program conducted by Bellcore and its LEC clients. Since 1986 this study, known as the Fiber Optic Cable System Field Failure Analysis (or Field Tracking Study), has been collecting field data on fiber optic cable damage. Detailed data on individual field failures is submitted to Bellcore (using its questionnaire) by outside plant personnel of the participating companies. Results of the Field Tracking Study summarizing the causes and type of fiber optic cable damage, cable repair information, and observed field failure rates have been published.

The study results, which provided direction for the Group's subsequent data request, indicated that the most frequent cause of fiber optic cable damage was damage by dig-up and that a substantial fraction of dig-ups occurred in spite of prior notification to the cable owner by the digging contractor.

Based on this data, the Group aimed its efforts at obtaining detailed root cause data on dig-ups, while continuing to collect data on all causes of cable failure. Companies that had not been participating in the Field Tracking Study received this questionnaire while regular study participants also received a questionnaire combining questions of historical relevance with questions of specific interest to the Focus Group.

A single questionnaire was to be completed for each fiber optic cable failure whether or not the damage affected customer service. For the Group's study, a failure is defined as any event which either 1) causes a fiber optic system to cease to operate to specifications, or 2) which requires timely maintenance activity.

In addition to receiving information specific to field failures, the Group requested data on the deployed population of fiber optic cable. Each company was asked to provide recent totals of their deployed fiber, sheath, and route mileage of fiber optic cable in their networks. For example, a one mile trench containing two 12-fiber cables is equivalent to one route mile, two sheath miles and 24 fiber miles. Mileage totals were classified according to installation (see para. 5.2.4, "Failure causes by installation," for definitions), estimates on cable mileage, and amount of fiber cable possessing certain attributes such as metallic armoring. This data was augmented by a recent FCC report on deployed fiber optic cable mileage.

This population data was used to make comparisons of failure susceptibility among various installations and to measure the effectiveness of certain damage prevention practices such as visible markers placed along cable routes. The questionnaire used for population data is shown in Appendix 2.

4.2 Data collection period.

The Focus Group requested data on FCC reportable field failures which occurred between April 1, 1992 and August 31, 1992, and data on all fiber cable failures (whether affecting service or not) occurring between September 1, 1992 and February 12, 1993. Some companies elected to submit failure reports on non-FCC reportable outages occurring prior to September 1, 1992. As of February 12, 1993, 160 failure reports were received in response to the Group's data request. These failures occurred between March 1, 1992 and February 4, 1993.

It should be noted that two major hurricanes, Andrew and Iniki, occurred during the data collection period. As evident in reports filed with the FCC concerning these storms, significant damage was incurred by aerial fiber optic cable in the affected areas. Individual reports describing cable damage caused by these hurricanes was not provided to the Group and consequently not included in the analysis. Therefore, summaries of failure data presented in Section 5 are not influenced by these unusual events.

4.3 Data collection procedure.

As mentioned, the Field Tracking Study is an ongoing data collection effort within Bellcore and its LEC clients, all of whom responded to the Focus Group's data request. Companies that had not been participating completed questionnaires and returned them to Bellcore's point of contact for NRC data collection. Many participating companies continued to submit failure reports directly to the Field Tracking Study director at Bellcore.

Procedures used to complete individual reports varied among companies. In some instances, failure reports were completed by field personnel who performed actual repair work. They then forwarded a report to Bellcore's central coordinator for data submission. In other instances, failure reports were completed by the company coordinator and based on field generated data. Each failure report included a contact person and telephone number to facilitate any necessary follow-up. The Field Tracking Study director is responsible for verifying data accuracy and completeness of all reports and if needed, for contacting reporting companies to clarify data.

4.4 Analytical methods.

Various analytical methods were used to arrive at the Focus Group's conclusions. Pareto analysis identified the most significant failure causes while statistical hypothesis testing was conducted to compare variations in failure cause, relative failure probability, and failure severity among different cable installations. To compare these variations, failure reports from companies which exhibited high levels of reporting were counted and compared with deployed mileage totals from those companies. The hypothesis testing was performed using a 95% confidence level (or 5% significance level).

4.5 Results of fiber optic cable data analysis.

The data analysis was conducted on 160 field failures reported to the Group as of February 12, 1993. In some instances, the analysis is augmented with recent data from the Filed Tracking Study. This section presents a summary of significant findings, information on failure causes, variations in relative probability among cable installations, and repair time statistics.

4.5.1 Summary of findings.

Following is a summary of the significant findings of the data analysis:

o Dig-ups were the single largest cause of reported fiber cable damage accounting for 58% of the reported failures.

o Failure of the excavator to provide prior notification was responsible for 33% of the reported dig-up failures.

o Of the reported dig-ups, 40% occurred in spite of prior notification by the excavator, accurate cable locates, and proper temporary marking of the sub-surface cable routes.

o Only one reported dig-up failure can be directly attributed to improper burial depth, however the correlation between burial depth and failure probability has not been determined.

o Aerial cable was damaged by a wide variety of causes.

o The causes of cable damage leading to large outages were not substantially different from causes of damage leading to outages of lesser impact.

o There was no significant difference in the relative failure probability between sub-surface and aerial fiber cables.

o There was no significant difference in the relative failure probability due to all failure causes between direct buried cables and cables installed in underground ducts.

o Differences were observed in the relative probability of failure due to dig-ups between direct buried and underground cable. These differences are probably influenced by differences in geography and may not solely be due to the presence of the duct structure.

o The mean time required to restore service after a cable failure is considerably less than the mean time required to completely repair a damaged cable.

5. Causes of Fiber Optic Cable Damage

This paper centers on fiber cable failures caused by "dig-ups" and the factors which contribute to and can best account for such incidents. Cable dig-ups typically occur during excavation work at construction sites such as business office complexes, housing developments, public works (e.g., roads, highways, malls, facilities such as sports arenas, airports and others). Cable dig-ups can also occur on private property (e.g., farms) or can result from excavating activities undocumented by federal, state, or county licensing authorities.

5.1 Detailed failure causes, definitions and root causes.

Through February 12, 1993, 160 failure reports were received in response to the Focus Group's data request. Nearly 100% of these failures resulted from a broken fiber or cable. This finding is consistent with results of Bellcore's Field Tracking Study which indicates that over 93% of cable failures reported between 1990 and 1992 resulted from a broken fiber or cable. Other types of failures which occur in the field include high loss induced by pressure on optical fibers (known as microbending), sheath damage, and procedural errors not resulting in facility damage.

There is a wide variety of external events reported to have caused fiber and other cables to fail and frequently a chain of events can lead up to cable failure. For example, a dig-up can occur because the digging contractor failed to provide prior notification. Therefore, in the data analysis presented here, failure causes have been categorized according to immediate causes and root causes. An immediate cause is the final event, such as a dig-up, which directly causes the failure. The root cause is the critical event, e.g., lack of proper notification, which eventually leads to a failure.

Table 1 lists the immediate causes of fiber optic cable failures reported to the Group. Every cause responsible for more than one reported failure is

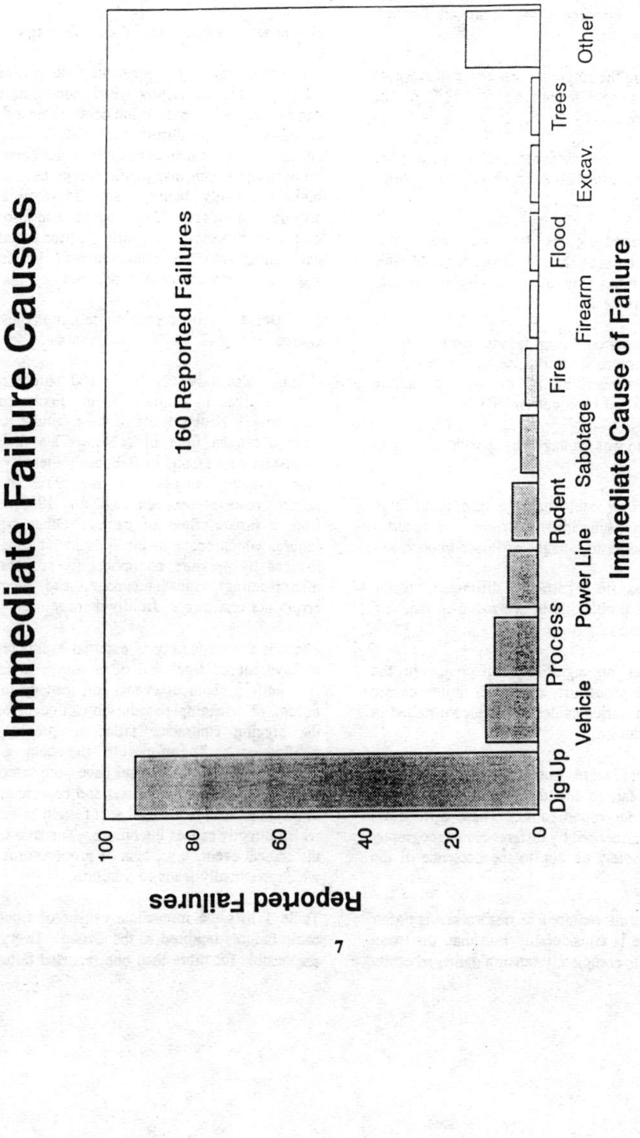
Immediate Failure Causes
160 Reported Failures
Reported Failures
100
80
60
40
20
0
Dig-Up
Vehicle
Process
Power Line
Rodent
Sabotage
Fire
Firearm
Flood
Excav.
Trees
Other
Immediate Cause of Failure

Figure 1

listed. Figure 1 provides comparisons of the numbers of these reported immediate failure causes.

Table 1. Immediate Causes of Fiber Optic Cable Failure

Cause	# of Reports	% of Total
Dig-ups	93	58.1
Vehicle	12	7.5
Process Error	11	6.9
Power Line	7	4.4
Rodent	6	3.8
Sabotage	4	2.5
Fire	3	1.9
Firearm	2	1.3
Flood	2	1.3
Excavation (non dig-up)	2	1.3
Fallen Trees	2	1.3
Other	16	10.0
Totals	160	100.0

As shown in Table 1, dig-ups are the single largest cause of fiber optic cable damage accounting for more than 58% of the reported failures. Following dig-ups, vehicle induced damage is responsible for 7.5% of the reported failures and process error comprises 7.0% of the reports. The relative magnitudes of these three failure causes, along with most of the remaining failure causes, correlates well with data previously presented in Bellcore's Field Tracking Study.

The following paragraphs define and/or describe each immediate cause and provide some general discussion about the reported failures.

Dig-ups

A dig-up is defined as damage to fiber optic cable during an activity to penetrate the ground. Dig-ups occur during sign placement, road grading, trenching, and sub-surface facility installation or repair. Occasionally dig-ups involve a procedural error such as failure to properly locate the sub-surface cable. For the purposes of this analysis, dig-ups resulting from procedural errors are classified as dig-ups. Historically, dig-ups have consistently been responsible for 55% to 60% of cable failures reported to the Field Tracking Study.

Of the 93 dig-up reports submitted to the Focus Group, 83 reports specify the excavator involved in the dig-up. In some instances the excavator's affiliation (e.g., telco contractor, highway dept. contractor) was not determined. However, 69 reports indicated whether or not the excavator was employed by the telephone company whose cable was damaged. In 87% of these cases, the excavator involved in the dig-up was not working on behalf of the facility owner while 13% of the reports indicate the dig-up was caused by the telephone company or telephone company contractor.

Other excavators involved in significant numbers of cable dig-ups include: electric, gas, and water company contractors, and highway department contractors. Private landowners account for only four of the 83 reports.

Vehicle Damage

Typically, this type of cable damage results from vehicle collisions with utility poles which support aerial cable. In three reports a passing vehicle caught a sagging aerial cable. One report involved a vehicle colliding with a highway overpass which housed fiber optic cable. Only one reported sub-surface cable failure was caused by a vehicle.

Process Error

Process error damage includes failures caused by telco personnel performing maintenance or installation work, but does not include procedural errors leading to dig-ups. Damage due to process error has consistently been responsible for about 7% to 8% of failures reported to the Field Tracking Study. The 11 process errors reported to the Group occurred under a wide variety of circumstances. Three failures were caused when an in-service fiber optic cable was accidentally severed during removal of an obsolete cable. Three failures occurred during the repair of a nearby damaged cable.

Power Line Contact

Damage caused by power lines occurs when an electric power cable makes contact with an aerial fiber optic cable or its metallic messenger strand. When electric current begins to flow in the messenger strand the resulting heat burns through the fiber cable. This type of cable failure was not widely reported to Bellcore's Field Tracking Study prior to 1992.

Rodent Damage

Damage caused by rodents affects aerial, sub-surface, and intra-building fiber cable and includes failures caused by gnawing animals such as mice, rats, gophers, and beavers.

Frequently, rodent induced damage results in only a partial failure of a cable. Only one of the six rodent induced failures reported severed every fiber in the cable. By contrast, over 80% of reported failures caused by dig-ups severed every fiber in the cable. Considering all rodent induced failures reported to the Field Tracking Study, 50% have resulted in the failure of every fiber in the cable.

Sabotage

Failures due to sabotage are caused by deliberate human activity whose sole purpose is to damage cables. Two of the four sabotage failures reported to the Focus Group were caused when a handhole (a buried concrete box containing a splicing point) was broken into.

Fire

Fire damage is defined as cable failure resulting from fire(s) to nearby structures and typically affects aerial and intra-building cable. In three reports submitted to the Focus Group, fires to nearby buildings and a building containing the damaged cable were described. These failures excluded cables which were set on fire upon contact with power lines.

Firearm

Firearm induced damage occurs when aerial cables are struck by a bullet or shotgun blast. Failures caused by firearms frequently do not sever the cable and can therefore be difficult to locate. These failures are not considered sabotage.

Floods

Floods cause damage when large quantities of water sever a fiber optic cable. Hydrogen permeation failures resulting in high loss and fiber fractures caused by the corrosive effects of small amounts of water have, in rare instances, been observed in the field but did not account for any failures reported to the Focus Group.

Excavation Damage

Excavation damage included two failures caused by human activity near an excavation site, but were not directly caused by any digging activity. One failure was caused by a large rock accidently dropped on an exposed concrete duct thereby breaking the duct and severing the cable. The second failure was caused when excavation machinery was used to exert a significant amount of downward pressure on the soil. This downward pressure was sufficient to sever a direct buried cable located beneath the machinery.

Falling Trees

In two reports, falling trees were considered the immediate cause of cable failure. These failures occurred when an aerial cable was severed by a falling tree limb. Falling tree limbs and sagging branches occasionally led to failures caused by power line contact or damage caused by vehicles.

Other Causes

Other causes of reported cable failure were mostly

extrinsic damage caused by human activity. These included failures caused by tree trimming, deliberate cable severing in emergency situations, and other unusual situations. This category also includes 10 failures where the causes were either not reported or not documented.

5.2 Fiber optic cable dig-ups.

A major emphasis of the Focus Group's data request was to gather detailed information on cable dig-ups. This involved determining the root causes of dig-ups and studying the effects of cable installation and damage prevention practices on reducing the probability of dig-ups.

5.2.1 Root causes of dig-ups.

Of the 93 cable dig-ups reported, 61 reports contained sufficient information to assign a probable root cause. Root causes are listed in Table 2 and presented graphically in Figure 2. It should be remembered that these failure reports were submitted by the facility owner, not the excavator. In at least one instance the circumstances surrounding the dig-up (as reported by the telephone company) have been disputed by the excavator.

Table 2. Root Causes of Fiber Optic Cable Dig-ups

Root Cause	# of Reports	% of Total
Digging Error	24	39.3
No Notification	20	32.8
Cable Unlocated	7	11.5
Inaccurate Locate	6	9.8
Incorrect Notification	3	4.9
Shallow Cable	1	1.6
Totals	61	100.0

As shown in Table 2, the leading root cause of reported cable dig-ups is excavator error(s) made during digging. These failures occurred after the facility owner was notified and the route properly located and marked. The second largest root cause is excavator failure to provide notification prior to digging. The following defines each root cause and discusses various details of the reported failures.

Digging Error

This root cause of cable damage is considered a digging error if: the contractor provided accurate prior notification, the route was accurately located and marked, and the cable was buried at a proper depth with sufficient clearance from other sub-surface structures. One failure where the contractor provided notification but did not wait for the cable to be located was also considered a digging error.

ANSI standard EIA/TIA-590, Standard for Physical Location and Protection of Below-Ground Fiber Optic Cable Plant, contains procedural recommendations for excavators to follow when digging near sub-surface cable. In particular, these procedures specify digging with hand tools within a "tolerance zone," specified at 18 in. from the edge of the facility (unless otherwise specified by local state or county laws).

Of the 24 dig-ups whose root cause was a digging error, only one report indicated that digging was performed with hand tools within the tolerance zone while 13 reports indicated that hand digging had not been performed at all.

Some companies require that a representative remain on-site to monitor an excavation. Of the 24 reported dig-ups where root cause was a digging error, three reports indicated that a representative of the telephone company was on-site while digging was performed while 14 reports indicated that a representative was not on-site.

No Notification

Ensuring that excavators provide consistent, accurate, and prior notification of excavation activities to the facility owner, or other responsible organization such as a "one-call" agency, is one of the principal lines of defense for preventing cable dig-ups.

84-088 0 - 94 - 6

Root Causes of Fiber Optic Cable Dig-Ups

61 Dig-Up Reports

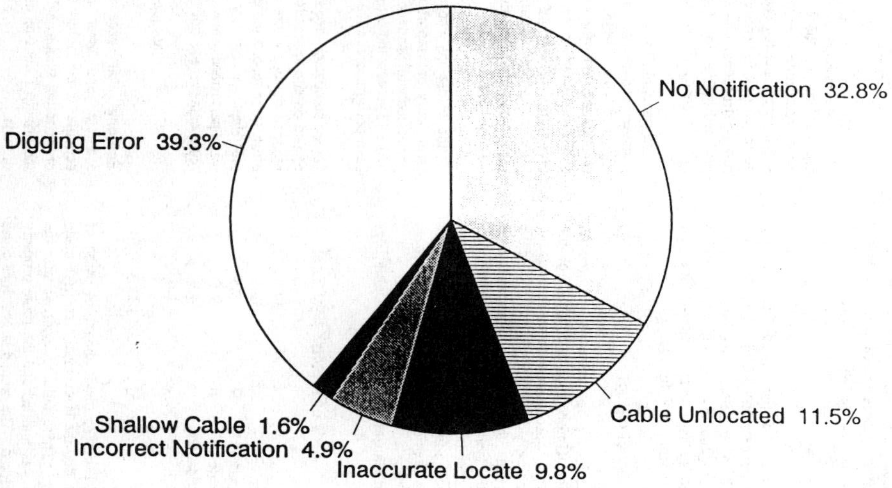

Figure 2

In 33% of the dig-ups having root causes, the excavator failed to provide any notification prior to digging. This finding is consistent with data in Bellcore's Field Tracking Study which indicates that the fraction of dig-ups occurring without prior notification has remained relatively constant over the past few years averaging 37% from 1989 through 1992.

Although a substantial fraction of the reported dig-ups occurred after notification was provided, prior notification appears largely to be acted on effectively by fiber optic cable owners. Two IXCs recorded an annualized total of over 3.5 million locate requests (prior notifications). These requests resulted in roughly 560,000 dispatches to locate sub-surface cable. Of these, only three fiber cable dig-ups occurred after the cable was located. Similar data comparing the number of dig-ups with the number of prior notifications was shared by one major LEC. This information indicates that excavators providing accurate, prior notification has been highly effective in reducing cable damage due to dig-ups.

Because of the success in avoiding dig-ups by acting upon prior notification, the lack of notification is considered to be the root cause of every dig-up in which prior notification was not provided.

Cable Unlocated

In these seven failures, prior notification was provided by the excavator but the facility owner or locating company failed to establish the presence of a cable which was then eventually damaged. In six of seven reported cases, locating personnel were dispatched in an attempt to locate a cable. In three of these six reports, the locating personnel properly located other nearby sub-surface cables. Of the six reports in which locating personnel were despatched, five locates were performed by a locating company while one locate was performed by the telephone company itself. In two of the failures it was reported that the damaged cable was not shown on locating company maps.

Inaccurate Locate

The root cause of these damage reports was considered an inaccurate locate because the cables' presence was determined but their locations were not accurately identified. For three of these six incidents, the cables were directly buried in soil while two incidents involved cables deployed in underground ducts. One report indicated a cable took an unexpected right-angle bend. The telephone company performed three of the locate attempts while two were performed by a locating company.

Incorrect Notification

In these cases, the contractor provided prior notification but did not accurately describe the location of the digging work to be performed. Therefore, the actual cable which was near the proposed excavation was never located.

Shallow Cable

The root cause of one failure was due to shallow cable burial depth. In this case, the route was properly located and marked and excavation was conducted with hand tools, but the cable was buried to a depth of only four inches. The cable was installed at this depth because the underground right-of-way was highly congested. This depth is substantially more shallow than industry standards recommend, e.g., EIA/TIA-590 specifies 30 in. as the minimum cover for interoffice cables and underground duct, 24 in. for feeder cables, and 18 in. for customer drop lines.

There were 63 dig-up reports received indicating the depth of a below-ground cable. Although four of these reports indicated cable depth at the point of failure was less than specified by the EIA standard, two failures occurred because no prior notification was given and one report did not provide sufficient detail to indicate a probable root cause.

5.2.2 Effect of cable depth on dig-up probability

As discussed in the previous section, shallow burial depth was identified as the root cause for only one reported dig-up. To fully quantify the relationship between burial depth and relative dig-up probability, comparisons must be made between the numbers of reported dig-ups at given depth and the quantities of sheath mileage deployed at those depths.

Figure 3, a histogram of burial depth for 63 of the reported dig-up failures, shows that most dug-up cables are buried at a depth of between three and five feet. However, since many companies specify a burial depth of three to four feet for fiber optic cable installation, this result is to be expected. Unfortunately, most companies do not maintain detailed information on the quantity of sheath mileage deployed at various depths. Thus, variations in reliability for different burial depths cannot currently be quantified. Bellcore is continuing to investigate the effect of burial depth on dig-up probability.

5.2.3 Effect of permanent marking on dig-up probability.

EIA/TIA-590 recommends use of permanent above ground markers and/or underground warning tape to identify the general location of a sub-surface cable. Permanent above ground marking is considered effective in reminding excavators to provide proper notification prior to digging. Most companies responding to the Focus Group's data request indicate their standard practice is to place visible permanent markers along sub-surface cable routes.

One way to quantify the impact of above ground markers on reducing the likelihood of dig-ups, is to compare dig-ups occurring (per sheath mile) in marked installations to those occurring in unmarked installations. However, this analysis is complicated by several factors including those shown below.

o Some companies mark nearly 100% of their sub-surface routes and, as expected, reported no failures along unmarked routes.

o Some companies do not keep precise mileage totals of marked and unmarked routes.

o Some companies indicated that although it is standard practice to mark 100% of their sub-surface routes, strict adherence to this practice is difficult in certain metropolitan areas.

o Metropolitan areas can have inherently better records for prior notification than rural areas.

o Although markers may have been placed during installation, the marker(s) nearest a proposed excavation may have been knocked down, broken, or become obscured by brush.

In attempting this analysis, data was considered from one company which exhibited good reporting while having a mixed population of marked and unmarked cable routes. This data was supplemented with data from the same company submitted to Bellcore's Field Tracking Study between 1991 and 1992. Table 3 summarizes these data.

Table 3. Reported Dig-ups per Sheath Mile - Marked and Unmarked Routes

	Marked Routes	Unmarked Routes
% of Reported Dig-ups	66.7%	33.3%
% of Sheath Mileage	63.6%	36.4%

Table 3 indicates that, for this company, there is virtually no difference in the reported dig-ups per sheath mile between marked and unmarked routes. When interpreting this result it should be remembered that:

1) much of the unmarked sub-surface cable is probably located in metropolitan areas, these areas can have inherently better notification records and therefore experience fewer dig-ups per deployed sheath mile than in rural areas,

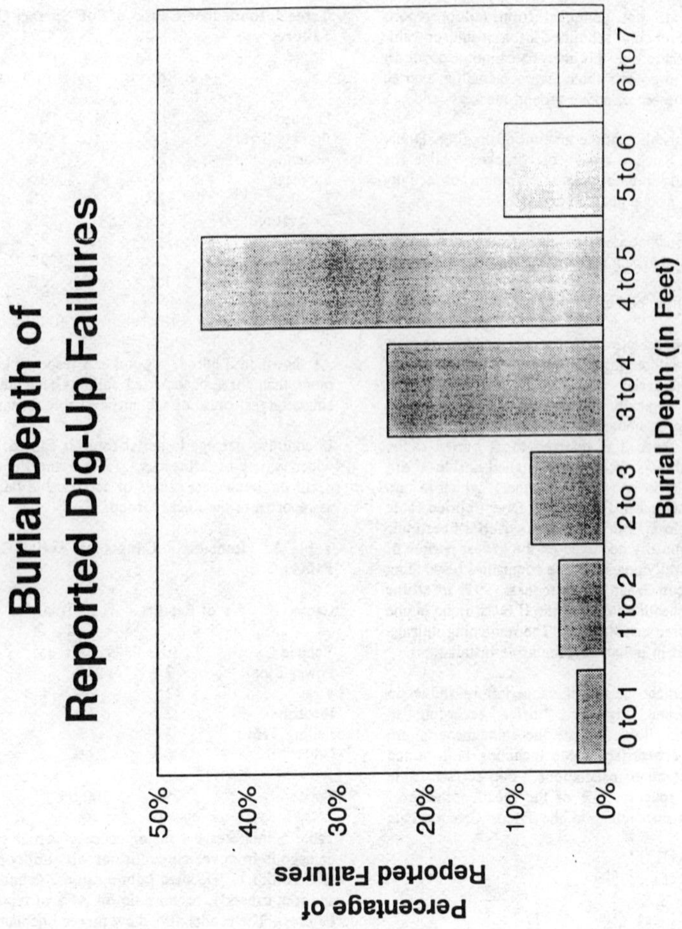
Burial Depth of
Reported Dig-Up Failures
Percentage of
Reported Failures
50%
40%
30%
20%
10%
0%
0 to 1
1 to 2
2 to 3
3 to 4
4 to 5
5 to 6
6 to 7
Burial Depth (in Feet)

Figure 3

2) this data was compiled from failure reports which were only submitted as a result of cable damage, therefore, this analysis cannot account for potential dig-ups which were actually averted through the use of above ground markers, and

3) it is possible that the amount of digging activity occurring near areas of marked cable is substantially higher than the amount of activity occurring near unmarked areas.

Bellcore is continuing to investigate the effect of permanent marking on reducing dig-up probability.

5.2.4 Failure causes by installation.

In the United States fiber optic cable is principally deployed in three installations: underground, direct buried, and aerial. Cable installed in a rigid duct structure, typically found under city streets, is referred to as underground cable. Cable directly placed in the soil is referred to as buried cable. Buried cable is frequently installed inside a soft innerduct (also known as ductliner), which in turn can be installed in PVC pipe. Direct buried cable is usually located in rural areas. Aerial fiber optic cable is typically deployed on the lowest portion of joint use utility poles. Of the companies responding to the Focus Group's data request, 99% of all the deployed sheath mileage in the U.S. is found in one of these three installations. The remaining mileage is deployed in indoor or submarine installations.

The immediate causes of cable failure shown in Table 1 and Figure 1 differ according to installation. Table 4 lists the immediate failure causes of sub-surface cable including both buried and underground installations. Sub-surface cable comprises roughly 75% of the sheath mileage of companies responding to the Focus Group's data request.

Table 4. Immediate Causes of Sub-Surface Cable Failures

Cause	# of Reports	% of Total
Dig-ups	93	71%
Process Error	11	8%
Rodent	5	4%
Sabotage	4	3%
Flood	2	2%
Excavation (non dig-up)	2	2%
Vehicle	2	2%
Other	12	9%
Totals	131	100.0%

As shown in Table 4, dig-ups are responsible for more than 70% of reported failures and are the single largest cause of sub-surface cable damage.

In contrast, damage to aerial cable is caused by a wider variety of influences. Table 5 and Figure 4 detail the immediate causes of aerial cable failures as reported to the Focus Group.

Table 5. Immediate Causes of Aerial Cable Failure

Cause	# of Reports	% of Total
Vehicle	10	34%
Power Line	7	24%
Fire	2	7%
Firearm	2	7%
Falling Tree	2	7%
Other	6	21%
Totals	29	100.0%

Table 5 indicates the major cause of aerial cable damage is from vehicle collisions with utility polls (see para. 5.1, "Detailed failure causes, definitions and root causes"), responsible for 34% of reported failures. The results also show power line damage is a significant cause of aerial cable failure

Immediate Causes of Aerial Fiber Optic Cable Damage

29 Aerial Cable Reports

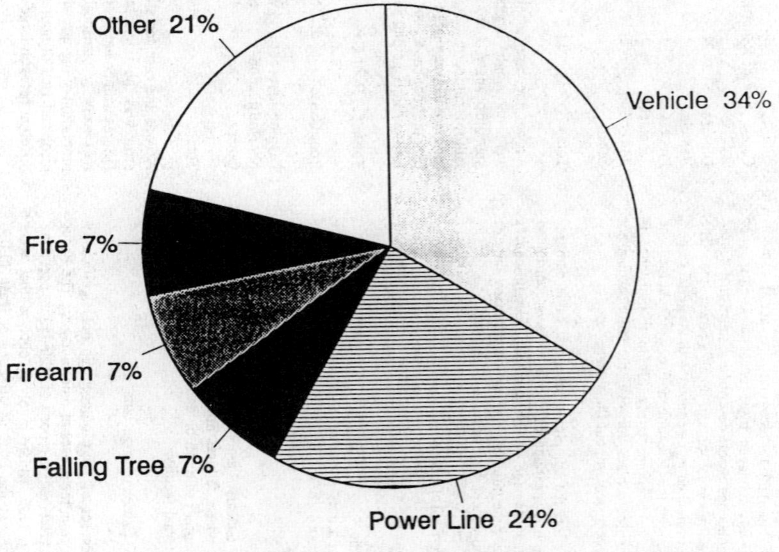

Figure 4

accounting for 24% of the reported failures.

It is important to recognize that aerial cable is frequently damaged by a series of events. For example, one report described a storm that caused a tree limb to fall on and break a power line. The power line then made contact with the aerial cable's messenger thus creating a fire which burned the cable. The immediate cause in this instance is power line contact, however the failure would not have occurred without the storm.

To reflect the complexity of aerial cable damage, Table 6 lists influences known to have contributed to such damage. The sum of these influences exceeds the number of reports since many failures are influenced by more than one event.

Table 6. Influences on Aerial Cable Damage

Cause	# of Reports	% of Total
Vehicle	10	34%
Power Line	8	24%
Weather	7	24%
Trees	7	24%
Fires	3	10%

Table 6 clearly demonstrates that unlike sub-surface cable whose failures are caused primarily by dig-ups, aerial cable can be damaged by a wide variety of external factors.

5.2.5 Causes of large outages.

The causes of large outages did not significantly differ from the causes of other outages. The Focus Group received 33 reports of cable failure occurring during 1992 which involved outages that were reported to the FCC as affecting at least 30,000 customers, special offices, and for a duration of more than 30 minutes. Table 7 lists the causes of these large outages.

Table 7. Causes of Large Outages

Cause	# of Reports	% of Total
Dig-ups	21	64%
Vehicle	3	9%
Process Error	3	9%
Sabotage	3	9%
Power Line	2	6%
Other	1	3%
Totals	33	100.0

As was the case for all reported failures, dig-ups were the single largest cause of large outages. Statistical analysis indicates that the proportion of failures caused by dig-ups was not significantly different between large and other outages. Table 8 lists the root causes for 19 of the 21 dig-ups which resulted in large outages.

Table 8. Root Causes of Large Outages

Root Cause	# of Reports	% of Total
Digging Error	9	47%
No Notification	6	32%
Others	4	21%
Totals	19	100.0

As was the case for all reported failures, digging error was the single largest root cause of dig-up failures with no notification being the second largest root cause. Statistical analysis reveals that the proportion of dig-ups whose root causes are digging error and lack of notification was not significantly different between large and other outages.

5.3 Comparisons of relative failure probability among installations.

To determine the relative failure susceptibility among installations, failure reports from the companies exhibiting consistent reporting were counted by installation and then compared with sheath mileage totals by installation from those companies (small variations in reporting periods across companies were ignored). Hypothesis tests were then conducted by comparing the reported failures per sheath mile for each installation. The comparisons were made at a 5% significance level.

5.3.1 Comparing aerial with sub-surface cable.

Although the failure causes for sub-surface cable are vastly different from those of aerial cable, the reported failure probability (measured in failures per sheath mile per year) is nearly identical for the two installations.

Six companies exhibited good reporting over this variable and provided deployed mileage totals for the two installations. Table 9 lists the percentage of reports received along with the percentage of sheath mileage deployed by these companies for each installation.

Table 9. Aerial and Sub-surface Cable - Relative Failure Probability

	Aerial	Sub-surface
% of Reported Failures	24%	76%
% of Sheath Mileage	25%	75%

As shown in Table 9, there was virtually no difference in the reported failure probability between aerial and sub-surface cables.

To further evaluate any possible differences between installations, similar data from Bellcore's Field Tracking Study was analyzed. Failures reported from four companies exhibiting consistent reporting between 1989 and 1991 were compared with sheath mileage totals from those companies. Using the same statistical test, this data again reveals that the observed failures per sheath mile per unit time are nearly identical for the two installations.

5.3.2 Comparing direct buried with underground cable.

When examining the difference in relative failure probability between buried and underground cable, it was necessary to combine data collected in response to the Focus Group's data request with data collected by Bellcore's Field Tracking Study in the first half of 1992.

Bellcore data was included in this analysis because some of the companies who reported consistently did not indicate whether damaged sub-surface cable was deployed in buried or underground installations. Further, no differentiation between buried and underground cable was made in some of the deployed mileage information.

When comparing the failure probabilities of buried and underground installations, it should be noted that buried and underground cables are deployed in geographically diverse regions. Therefore, any observed difference in relative failure probability is strongly influenced by regional geographic features and probably not due solely to the presence of the duct structure itself.

Upon analyzing the data, it was found that when considering all failure causes, there was no statistically significant difference in the reported failure probability between buried and underground installations. However, it is likely that some difference in failure probability exists, but more data is needed to verify whether this is the case. Further, when considering only failures due to dig-ups, a statistically significant difference was observed.

Six companies exhibited good reporting over this variable throughout 1992. For each installation, Table 10 lists the percentage of failure reports received along with the percentage of sheath mileage deployed by these companies.

Table 10. Buried and Underground Cable - Relative Failure Probability

	Buried	Underground
% of Reported Sub-surface Failures	67%	33%
% of Sub-surface Sheath Mileage	54%	46%

Statistical analysis of this data reveals that there was no significant difference in the reported failure probability between direct buried and underground cables.

A somewhat different result is obtained by only considering failures caused by dig-ups. Table 11 lists the percentage of failure reports received along with the percentage of sheath mileage deployed by these companies in each installation.

Table 11. Relative Failure Probability Due to Dig-ups

	Buried	Underground
% of Reported Failures	71%	29%
% of Sheath Mileage	54%	46%

Statistical analysis of this data reveals that there is a significant difference in the reported failure probability between the two types of installations.

It is suspected that geographic influences play a significant role in accounting for this difference. In general, most companies stated that underground cable is deployed in urban areas while direct buried cable is prominent in rural areas. Sheath mileage data on the installation of deployed cable by urban and rural environments was not available in sufficient detail to be included in the analysis.

However, between 1986 and 1991 Bellcore's Field Tracking Study collected data on whether fiber optic cable failures occurred in urban or rural areas as a function of the failed cable's installation type. This information can then be used to assess the link between geography and failed cable plant. Table 12 lists, by geography, the number of failure reports received for buried and underground installations.

Table 12. Geography and Installation of Damaged Sub-surface Cables 1989 - 1990

Geography	Buried	Underground
Urban	15	23
Rural	40	6

Statistical analysis indicates a strong association between geography and the installation of failed cables. This provides evidence to suggest that the difference in failure probability due to dig-ups between buried and underground cable may be strongly influenced by geographic considerations.

To help determine the effect of duct structure in preventing cable damage, it is interesting to examine the severity of the damage caused by dig-ups to buried and underground fiber optic cable. One measure of damage severity to consider is the number of fibers broken as a fraction of the total number of fibers in the cable or the fraction of reports indicating a complete fiber break. Table 13 lists the average percentage of fibers broken during dig-ups for buried and underground installations and the percentage of reports indicating a complete cable break. This data combines data submitted to the Focus Group with data collected by Bellcore since 1989.

Table 13. Severity of Dig-ups Damage by Installation

	Buried	Underground
Average % Broken Fibers	95%	87%
% Complete Breaks	90.5%	83.0%

This data indicates that for dig-ups which do occur, the difference in the severity of damage incurred between buried and underground cables is not substantial. It should also be remembered that these events are not reported if there is no damage to the cable. Therefore, if the presence of a duct structure was instrumental in averting a possible dig-up, this would not be reflected in the data.

5.3.3 Comparing public with private right-of-way.

Data submitted to the Group contained information on the right-of-way of the damaged cable. Some companies provided deployed cable mileage totals in public and private rights-of-way. Public right-of-way includes highway right-of-way and other government property while examples of private right-of-way include property owned by railroads, power companies, and private landowners.

Of the IXCs responding to the Focus Group's data request, 75% of their fiber optic cable is deployed in private right-of-way. Of the LECs who provided this population data, approximately 20% of their fiber mileage is deployed in private right-of-way.

The immediate causes of reported cable failure did not vary substantially between cable deployed in public and private rights-of-way. The two most significant causes reported for these two installations were nearly identical. Table 14 highlights these causes.

Table 14. Failure Causes by Right-of-Way

Causes	Private	Public
Dig-ups	64%	71%
Vehicles	11%	7%
Others	25%	22%

The data also indicated that for failures caused by dig-ups, the fractions of dig-ups which occurred without prior notification were identical for cable deployed in each right-of-way.

To compare the relative failure probabilities due to all failure causes of cable deployed in public and private rights-of-way, reports from four companies exhibiting good reporting and who provided sheath mileage totals were counted. These companies include both LEC and IXC representatives. Table 15 lists the percentage of failure reports received along with the percentage of sheath mileage deployed by these companies in each installation.

Table 15. Public and Private Right of Way - Relative Failure Probability

Causes	Private	Public
% of Reported Failures	55%	45%
% of Sheath Mileage	67%	33%

Statistical analysis reveals that there was no significant difference in the reported failures per sheath mile between cable deployed in public and private rights-of-way.

5.4 Repair data.

Although the Focus Group did not investigate service restoration or cable repair strategies, data was collected on the time required to restore service after a service affecting cable failure, and the time required to permanently repair the damaged cable. Restoring service can be accomplished by rerouting traffic, temporarily repairing the cable, or by switching to a protection fiber. Complete cable repair includes: finding the failure location, splicing the failed cables or splicing in a new section of cable (or possibly replacing the entire cable run), and protecting the respliced area. Due to unusual conditions, long delays are occasionally experienced in complete cable repair. For this reason customer service is usually restored well before a cable is fully repaired.

Table 16 summarizes the mean, median, maximum, and minimum reported service restoration times and complete facility repair times. Histograms of service restoration time and cable repair time are provided in Figures 5 and 6 respectively. For convenience, repair times greater than 25 hours have not been included in Figure 6. However, these reports were included when computing repair time statistics.

Table 16. Summary of Repair Time Statistics

Statistic	Service Restoration Time (hours)	Complete Cable Repair Time (hours)
Mean	5.2	14.2
Median	4.6	10.0
Maximum	21.43	97.5
Minimum	.1	.48
# of Reports	132	66

As shown in Table 16, the mean time to completely repair a failed cable is slightly over 14 hours while the mean time to restore service is slightly over five hours. The complete repair time data does include five damage reports describing failures which were not service-affecting. However, the mean time to repair these failures was less than 15 hours and therefore has only a small affect on the statistics.

It is evident from Table 16 that the time required to restore service after a fiber optic cable failure is in general substantially less than the time required to completely repair the damage. It is also evident that the complete cable repair time distribution is skewed by the occasional long repair times. This skewness results in the four hour difference between the mean (average value) and the median (50th percentile) of the distribution.

6. Key Lessons Learned and Recommended Best Practices

The "best practices" definition as used in network reliability focus area technical papers, and here as well, is: best practices are those countermeasures (but not the only countermeasures) which go furthest in eliminating the root cause(s) of outages. None of the practices are construed to be mandatory; however, a very small number of countermeasures that are deemed by the Focus Team, and concurred by NOREST, to be especially effective countermeasures will be designated as "recommended." Service providers and suppliers are strongly encouraged to study and assess the applicability of all countermeasures for implementation in their companies and products, respectively. It is understood that all countermeasures, including those designated as "recommended," may not be applied universally.

6.1 Key lessons learned, best practices, new approaches.

Dig-ups are the single largest immediate cause of fiber optic cable failures (see para. 5.1, "Detailed failure causes, definitions and root causes"). In the prevention of such failures, industry-wide adherence to established processes and practices is required. However, universal adoption of these practices is complicated by the evolution of fiber optics and its associated implementation requirements. However, effective preventive practices exist and are implemented in the industry. These practices were identified through the examination of the damage prevention measures used, or absent, in the circumstances surrounding failures reported to the Focus Group's data request (see Fig. 2, "Root Causes of Fiber Optic Cable Dig-ups").

The Focus Group identified existing practices that if complied with by telecommunications companies

Distribution of Service Restoration Times

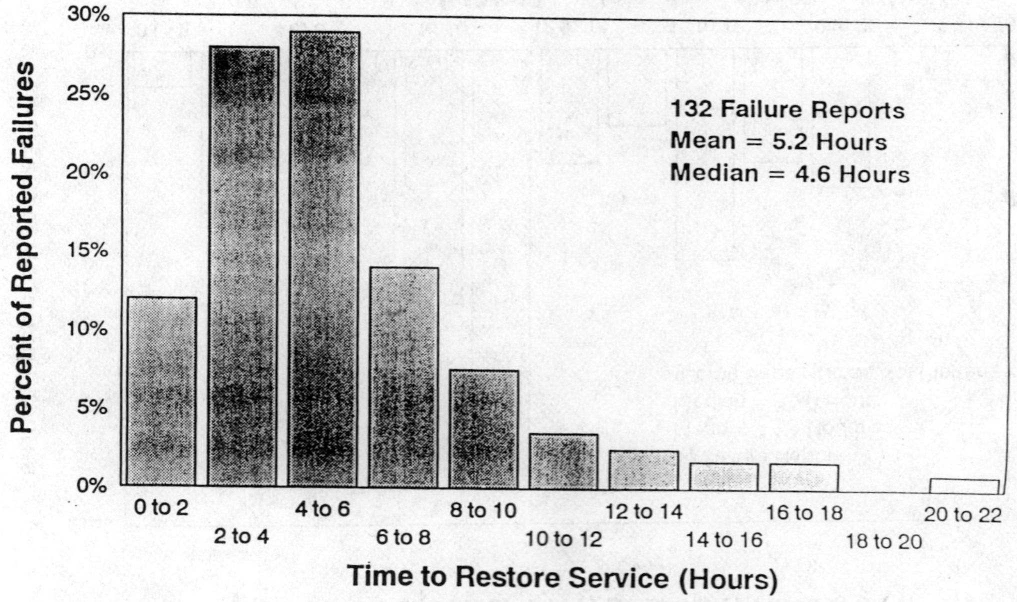

Figure 5

Fiber Optic Cable
Complete Repair Time Distribution

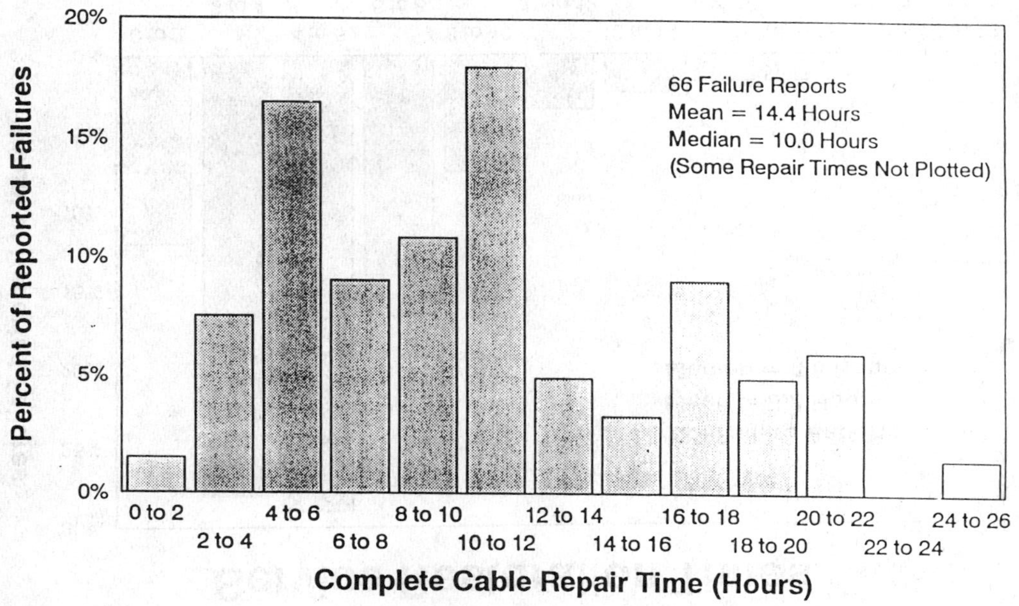

Figure 6

and excavators, will eliminate, mitigate, or at least minimize cable failures. Many of these practices and procedures relate to cable burial depths and cable mining operations, they are somewhat new and evolving while being implemented. In addition to identifying existing strategies that minimize fiber cable damage, jointly published ANSI, EIA, and TIA documents provide a good starting point for assessment of standards and procedures applicable to reducing the probability of fiber cable dig-ups.

The Group is also confident that reduced cable failures and outages will result if preventive procedures and practices are emphasized in personnel training as well as through elevating public awareness of the significance of cable damage and cable failure.

6.1.1 Best practices to prevent fiber cable damage caused by digging.

The analysis of cable failure information provided to the Focus Group's data request led the Group to align preventive practices, guidelines and procedures (by causes of fiber cable failures), damage caused by digging, and damage caused other than by digging.

The best practices to prevent fiber cable failures are further grouped by installation, maintenance and preventive sub-processes, they are:

o Engineering and construction

Compliance with industry and established company practices based on ANSI, EIA, and TIA minimum standards, e.g., burying cable at standard depths and distances from adjoining structures and marking cable routes with permanent markings and/or warning tapes.

o Call-before-you-dig

Notification to facility owners for cable location marking, enactment and enforcement of federal and state "call-before-you-dig" legislation and underground facility damage prevention laws.

o Locate the cable

It is the facility owner's responsibility to respond to cable locate requests and to promptly and accurately establish and mark cable route locations.

o Use additional appropriate prevention measures

Special care, consideration, and training to ensure fiber optic cable security, e.g., hand digging around safety zones, cooperation and communication between excavators and facility owners, technical training, general awareness elevation, and procedural training.

Complying with best practices and standards in combination with the above processes is essential to improving fiber cable security and thus reliability. It should be noted however, that elements of some practices are necessarily fixed and may conflict. Therefore, it is imperative that environmental, application, and economic considerations be incorporated in the design, implementation, and maintenance of fiber optic cable facilities.

6.1.2 Best practices to prevent fiber cable damage caused by other than digging.

The immediate causes of fiber optic cable damage examined in this paper, as reflected in the data request (see para. 5.1, "Detailed failure causes, definitions and root causes"), and other than by hand digging, are shown in Table 1 and Figure 1.

Associated with immediate causes are root causes (see para. 5.2.1, "Root causes of dig-ups") which, it is believed, could be dramatically reduced or even eliminated through consistent industry-wide observance of best practices.

Appendix 3, "Existing Practices," provides details of several current standards and procedures, not written by EC or IXC companies, aimed at preventing fiber cable damage. Included are ANSI, EIA, TIA, and NCS regulations.

6.1.3 Details - key lessons learned and best practices.

The Group identified the following baseline of best practices/counter measures as a representative means to bring about reductions in fiber cable failures including dig-ups. Many companies have adopted these practices and others should consider them where applicable. The best practices/counter measures encompass the following:

Adherence to Procedures - comply with, utilize, and benefit from existing standards and procedures.

Warning Tape - place tape 12 in. above the cable, do not deviate more than 18 in. from the outside edge of the facility.

Visible Cable Marking - place permanent markers at line-of-sight intervals to clearly indicate the placement and direction of cables.

Call-Before-You-Dig Legislation - enforce, enact, and/or revise state and federal underground facility damage prevention laws.

Respond To Locate Requests - execute prompt responses to contractor calls for cable locates.

Accurate Locates - expand locate personnel training and skill levels, quality control all work.

Enhanced Locating Equipment -utilize current, and/or emerging technologies with long range locating equipment, add a cable-unique sense tone carrier on each cable.

Use Of Plant Route Maps - secondary checking of plant drawings relative to marking.

Hand Dig In Safety Zone - perform hand digging exclusively within a two foot safety zone around the cable, and pot hole every 50 ft. for construction that is parallel to the cable route.

Technician Supervision - assign technical personnel to observe activities at work sites where digging is underway.

On-Line Technical Support - centralized support for technicians performing locates, timely answers to technical questions, facility drawing acquisition.

Cooperation With Contractors - easy access, open communications and flexible scheduling between facility owners and contractors.

Training - continuous refresher training and adherence to standards and procedures, train personnel to recognize conditions potentially hazardous to fiber optic cable.

Contractor Awareness - offer public service seminars, publish literature, and announcements by facility owners to educate contractors, right-of-way owners, and private property owners.

Contact With Land Owners - proactively educate and communicate with right-of-way owners to stress the importance of cable protection and security.

Patrol Cable Routes - regularly patrol fiber optic cable routes.

Audits/Surveys of Plant - periodically check and validate outside plant records and data.

Barriers - place barriers around poles and above ground structures.

Buried Cable - bury fiber cable in accordance with standards, mark and accurately document locations.

Buried Facilities - bury structures out of sight and to depths which are standards compliant.

Shielding - properly shield/insulate fiber cables.

Protective Devices - use rodent devices on poles and cable sheaths in rodent infested areas.

Stronger Conduit - use reinforced PVC pipe in rodent infested areas.

Separate Pole Lines - avoid joint use utility poles with fiber optic cable if justified by cost/benefit considerations.

No Visible Markings - avoid use of visible markings in areas prone to vandalism.

Secured Manholes - utilize lockable manhole covers in areas prone to vandalism.

Ventilate Manholes - install automatic purging devices in contaminated manholes.

6.1.4 Effective Countermeasures.

To varying degrees, some Focus Group member organizations have seen that adoption of the following practices has demonstrated positive results. These organizations find improved cable dig-up prevention results when expanding certain practices. Each of these, presented below, indicates what the practice is or should be in order to optimally address industry-wide concerns and issues focused on reducing cable dig-ups. The remaining practices, in addition to those detailed here, should be considered as candidates for close study and evaluation as part of a benchmarking effort.

Warning Tape - Warning tape should be used with buried installation but should not be relied upon as a primary means of locating cables. It should be characterized by its compliance with existing or emerging standards. The tape should be marked in recognizable and standardized format to convey its primary message. Warning tape alerts should be simple yet convey a high degree of message impact. "WARNING -- Optical Cable," or "STOP," printed on the tape will likely be very effective in protecting cable or other sub-surface facility assets.

The tape should be made of a non-metallic material to eliminate possible interference with electronic locating devices.

Technical Supervision/Plant Protection - It may at times be necessary to have qualified observer personnel on-site to provide cable plant protection. This protection is dependent on the size of the transmission cable, the nature of traffic on the cable, and other environmental factors. Activities having the potential to damage a cable should be determined well in advance and observer coverage should then be required and scheduled. Open excavations exposing cable/splices must also be guarded and protected during certain work activities. Stakes, flags, or other devices designed to capture the attention of construction contractors should be placed directly over a cable to show its direction, changes in direction, curves, and variations in depth (if any). Examples of cable plant protection and safety practices include:

1) accurately locating a cable,

2) digging a test pit to visually confirm a cable's location,

3) permanently marking a cable's location and its hand-dig-only-zone, and

4) guarding the cable plant when/if appropriate.

Effective cable plant safety procedures must be promoted by facility owners and observed, enforced and practiced at all times.

Hand-Dig-Only-Zone - A hand-dig-only-zone should be marked with lines or orange paint on each side of the cable. The zone is measured from the actual cable location which is best determined visually and whenever possible.

Size of the Hand-Dig-Only-Zone - The following conditions affect the width of a hand-dig-only-zone:

1) local legislation (some states have laws governing one-call systems and also specify the features of a hand-dig-only-zone, e.g., width and depth),

2) changes in cable direction,

3) buried cable slack loops,

4) cable branching stubs and offsets, and

5) other utilities crossing or paralleling a cable.

Test Pit Location - Whenever possible, network personnel should obtain an accurate cable location and determine depth by hand digging a test pit and visually confirming an electronic facility location. Test pits should be dug at intervals determined by the nature of the work activities being performed.

Training - Training programs, regularly upgraded in response to technology evolution, will sustain the various personnel skills required for fiber optic cable maintenance and operational restoration. All management and technical personnel must obtain the necessary knowledge and skills to satisfy organizational quality goals and customer service dependability expectations. The range of basic skills maintenance personnel should have include the ability to:

- describe and demonstrate proper procedures to purge and ventilate a manhole
- describe, demonstrate, and perform proper cable protection procedures
- use proper excavation methods in the area of a cable
- apply and use cable location techniques

Cable Locating Equipment - Low frequency locators operate in the 1 kHz, or lower, range and can only apply the signal by conduction for use in cable locating. The 60 Hz magnetic field of any cable with power on it interferes with the trace tone generated field. Each locator has its own unique electronic characteristics and practice is needed to learn how each device reacts to various situations. Electronic depth measurement is helpful, but it should not be used to direct powered excavation activities without first verifying location and depth by using hand dug test pits. Shown below are some of the factors that can influence the accuracy of electronic locator devices.

- spurious electrical fields
- other underground utilities
- operator or test set error
- improper ground connection
- soil acidity/soil moisture
- shield wires/cables
- variations in conductor depth
- transmitter proximity to receiver
- cable bonding
- cable deeper than other conductors
- cable not insulated
- cable diameter
- ground faults

Right-of-Way Maintenance - Proper right-of-way maintenance by facility owners communicates a strong message to others that underground plant is present in the right-of-way and that its safety and protection are guarded by the owners' enforced and practiced company procedures. Proper maintenance also allows for expeditious access to a fault location in the event of cable damage or failure.

Cable Patrol - To reduce the potential for damage to cable facilities, consideration may be given to implementing different levels of surveillance. New conditions arise between contact visits, construction plans and schedules change, and unforeseen activities by foreign parties can occur. The frequency of patrols may be established for each section of cable compliant with local conditions. As the levels of anticipated or planned activity increase, this may require differing frequencies of patrols needed to ensure adequate cable plant security.

Contractor Awareness/Prevention - Cable dig-ups by contractors are, historically, the number one cause of cable failures resulting in service outages. In a large number of these occurrences advance notification of the contractor's planned digging was not provided to the facility owner(s). However, given the opportunity to respond to contractor notification, owners can carry out an action plan to protect or guard their cable plant assets until hazards are removed. If facility owners are unaware of excavating activities in proximity to a cable right-of-way, or fail to learn of such activity, the results can be catastrophic. It is therefore imperative that facility owners strive to educate excavators, contractors, and the public at large about call-before-you-dig (one-call) procedures and legislation and shift the emphasis from damage repair to damage prevention. In support of this approach, some facility owners have designated personnel to cover a geographic region to actively promote call-before-you-dig damage prevention programs. These efforts should focus on and will be most informative to the following groups:

- small and large contractors
- private contractors
- utility owners
- federal, state, and local governments
- municipalities
- farmers and ranchers

For these programs to achieve optimum beneficial results, cable plant protection concepts and procedures should be presented to audiences with representatives from all or some of these groups. Additionally, an attempt should be made to develop joint efforts with other utilities and the one-call centers and their memberships to promote these programs and concepts. Some of the activities which will be integral to this program will include, but not be limited to, the following activities.

- contractor awareness nights
- contractor safety meetings
- proactive participation/board memberships in local one-call organizations
- damage prevention awareness booths at events such as:
    - utility/heavy equipment expositions
    - state farm bureaus
    - agricultural fairs
    - APWA equipment shows
    - county fairs
    - homebuilders associations

and, memberships in and/or program participation with trade groups such as:

- Associated General Contractors
- National Utility Contractors Association
- ULCC
- APWA

Facility owners may want to set priorities and concentrate their program efforts in areas where cable assets are at the greatest risk. These include areas of high capacity, heavy construction activity, shared rights-of-way, multiple underground facility crossings, areas having a high incidence of cable outages, or near misses. Cable routes defined as critical to facility owners should receive special attention.

Contact With Land/Property Owners - Communicating with land and property owners is another measure which can help bring about improved security for cable plant assets. These contacts should be scheduled to correspond with anticipated, planned or scheduled construction activities at or near cable or other sub-surface facilities on property owned by the persons contacted. Contact sessions with property owners could include inquiries concerning a variety of issues ranging from physical attributes and conditions of the property to changes in property ownership.

Warning Sign/Markers - Placing markers and signs will maximize plant protection by clearly delineating the presence and path of sub-surface cable routes while also reminding the public that advance notice must be given prior to excavating in the vicinity of marked cable routes. Markers should be placed at line-of-sight intervals to ensure route direction is clearly visible. Adjacent markers should be visible each way from an adjoining marker but should be no more than, e.g., 1000 ft. distant if land features permit such spacing. However, consideration must be given to areas prone to vandalism. Placement of warning signs and markers in these areas may not be desirable.

7. Metrics

7.1 Metrics to measure the effectiveness of solution recommendations.

Benchmarking conducted for the TBIC will define metrics for dissemination through the ECSA.

8. Path Forward

8.1 Benchmarking.

Best practices provide a certain level of assurance in cable damage prevention (see para. 6.1.3, "Details - key lessons learned and best practices"), yet cable damage continues to occur. To define additional practices, a benchmarking study will help identify strategies with the greatest potential for minimizing fiber cable failures. The scope of the study should include "world class" companies, within and external to the telecommunications industry, having excellent records of preventing service outages similar to and including fiber cable failures. The study should include a screening-out process to identify the most effective policies, strategies, and methodologies to implement for maintaining and protecting fiber cable assets. Specific areas a benchmarking study should cover include:

o PREVENTIVE/PROTECTIVE

- Call-before-you-dig
- One-call centers
- Contractor/customer education
- Alarming methods
- Marking and paroling

o INSTALLATION

- Cable design
- Construction methods
- Permanent marking techniques
- Documentation, record keeping

o MAINTENANCE

- Dispatch procedures
- Markers/marking techniques
- Mark and walk procedures
- Service restoration strategies

The Focus Group recommends a formal benchmarking study and solicits the TIBC to perform this task. It is very likely that the findings of an independent benchmarking effort will stimulate process and procedural improvements to increase fiber cable reliability.

MCI will manage the benchmarking study project through the TIBC and upon its completion submit study findings to the ECSA for distribution to all interested parties. The Fiber Cable Focus Group member companies have agreed to fund their portion of the study. Other NRC members have also agreed to join in the TIBC funding of efforts.

The process to perform the study is for the TIBC to support an independent third party consultant to "Identify 'Best In Class' business processes which when implemented, will lead member companies to exceptional performance as perceived by customers." This mission goal fully supports and complements the Group's goal, i.e., improving fiber cable reliability through improving its safety and security. The benchmarking process can access and evaluate performance data from non-industry companies thus providing an alternate, more universal perspective of best practices selectively applicable to minimizing fiber cable failures caused by dig-ups. The issues below would comprise a strong baseline from which a benchmarking effort could be launched, they include:

- which states have one-call systems in place
- what is the geographical coverage of each
- hours of coverage
- legislative details
- notification and response requirements
- excavation guidelines
- participants, members
- penalties for noncompliance

- exemptions

From a "Path Forward" perspective, Appendix 5, "Network Reliability Industry Initiatives," presents the activities of several focus areas.

9. Conclusions

After study data was analyzed and interpreted, several factors contributing to the cable failure problem became very clear but the dominant factor remains dig-ups (see para. 6.1, "Key lessons learned, best practices, new approaches," and Fig. 1, "Immediate Failure Causes"). The Focus Group then recognized that to meet near term needs a triad of actions will collectively, and adequately, address the persistence of this industry-wide problem. However, within the triad is a benchmarking recommendation. This activity will be required to assess the effectiveness of current problem solutions and, perhaps more important, the merit of new, yet to be defined long term solutions. The Group therefore recommends that this triad of activities be broadly implemented as an overall strategy for the industry to adopt until benchmarking yields new and/or revised problem resolution approaches and recommendations. The Group is confident that by implementing this triad, the industry will experience a dramatic reduction in current levels of fiber cable failures and their resultant service outages. The triad's primary components are presented below.

9.1 Call-before-you-dig as SOP (Standard Operating Procedure).

The Focus Group found that a very high percentage of reported fiber cable failures were caused by dig-ups (see Fig. 1, "Immediate Failure Causes") and that root causes such as no notification and digging error accounted for a large fraction of these dig-ups (see Fig. 2, "Root Causes of Fiber Optic Cable Dig-Ups").

An assessment of call-before-you-dig legislation revealed that the laws are inconsistent across the country and seem to lack the depth of enforcement and penalties required to deter excavators from digging without providing notification. Similarly, inconsistencies were identified in penalties imposed for digging errors after cables were properly marked. The Group therefore recommends more enforcement, expanded enactment, and/or revision of federal and state underground facility damage prevention laws to achieve enhanced uniformity of use. The legislative process must emphasize and include two key points, they are:

1) that any person or entity, e.g., municipalities and departments of transportation, excavating on utility rights-of-way must notify all underground utility owners at least two days prior to the time excavating work commences, and

2) that any person or entity, e.g., municipalities and departments of transportation, failing to provide notification within the prescribed time and causing damage to underground facilities, or causing damage to properly marked facilities, is financially liable. Such liability includes full compensation to the affected utility(s) for all costs of emergency restoration and repair work needed to return the facility to its operational condition prior to excavation. This is to include fines and penalties commensurate with state and federal regulations.

Several states have existing legislation that reflects these two points. For example, Pennsylvania requires a contractor to notify the underground facility owner "not less than three nor more than ten working days prior" to excavation. Virginia and Missouri have similar wording in their laws, except that minimum time is two full working days. These states also address the critically important issue of liability which contributes significantly to the credibility and validity of this legislation.

In Pennsylvania, the contractor is deemed liable for damage even if notification was provided but damage was caused by a failure to employ prudent methods (which can include hand-dug test holes). The legislation also calls for split liability if there was non-compliance by the contractor in addition to mismarking by the utility owner. Missouri has similar wording which states that compliance with the legislation's provisions does not excuse the contractor from "excavating in a careful and prudent manner." If cable damage occurs and the contractor had not complied with Missouri's legislation, the failure "shall be a rebuttable presumption of negligence."

Appendix 6, "Call-Before-You-Dig - Statute Analysis by State," presents a summary of state-by-state statutes (with effectiveness rankings). In order to make such call-before-you-dig legislation more effective, Appendix 7, "Model Legislation," provides an example of "model wording" for legislation which, if enacted, could produce the desired effects.

9.2 Best practices

The study suggests that if best practices (see para. 6.1.3, "Details - key lessons learned and best practices") are more uniformly implemented and enforced, then a measurable reduction in fiber optic cable damage will quickly result. Consequently, while the facility owners/operators interests are served well by existing best practices, industry failure to adopt and enforce these best practices can keep cable failures due to dig-ups a persistent industry-wide problem.

9.3 Benchmarking.

Benchmarking will help to isolate practices and procedures that are not realizing their full potential to protect facilities and also determine why they are failing to achieve expected results. Benchmarking will also bring about innovative approaches to facility security.

Once a benchmarking study is delivered to ECSA, industry members will have the opportunity to review and adopt its inherent recommendations as additions to present operating practices.

10. Acknowledgements

1. Mr. Samuel V. Lisle, Bellcore

Mr. Richard M. Fagerstrom, Bellcore

with references -

a. S.V. Lisle, "The History, Prevention, and Impact of Fiber Optic Cable Failures," to be published in the Proceedings of the National Fiber Optic Engineers Conference, San Antonio, TX, June 1993.

b. V.T. Hou, "Update on Interim Results of Fiber Optic System Field Failure Analysis," Proceedings of the National Fiber Optic Engineers Conference, Nashville, TN, April 1991.

c. J.M. Kraushaar, "Fiber Deployment Update - End of Year 1991," Federal Communications Commission, March 1992.

d. J.P Kilmer, et. al., "Post-Mortem Analysis of Optical Fibers and Cables," in Proceedings of the National Fiber Optic Engineers Conference, Nashville, TN, April 1991.

e. ANSI/EIA/TIA-590-1991, "Standard for Physical Location and Protection of Below-Ground Fiber Optic Cable Plant," July 1991.

2. Mr. Mike Michalczyk, Southwestern Bell Telephone Company

3. Mr. Paul E. Devaney, Applied Concepts, Inc.

11. References

None

12. Figures and Exhibits

Figures

Figure 1. Immediate Failure Causes

Mr. SHARP. Thank you very much, Mr. DeVaney. Let me quickly ask Mr. Huriaux, if I understood your testimony, some have suggested the value of having a uniform number, not necessarily having it go into one switchboard obviously, but it is used in the sense 911 is used, or 411, or those kind of things. If I understand correctly, your argument is that a lot has been invested in public education on specific numbers in specific areas and your concern is that is in a sense lost in public education. Am I correct?

Mr. HURIAUX. Yes, Mr. Chairman, that is correct. If we were starting today with no one-call systems in the country, then I think it would make sense to adopt a single number for use throughout the country.

But as I said in my testimony, we are 80 percent down the road to one-call implementation and at this point there has been a lot invested in public awareness of local numbers such as the Miss Utility number here in the Washington, Virginia, Maryland area. I think it would be hazardous to dump all the investment, the successful investment that has been made in public education in these programs just to have a national one-call number.

However, a clearinghouse, a national clearinghouse number that could be called by contractors who perhaps might want to be working in an area that they weren't familiar with to make sure that they were put in contact with the correct one-call center, I think would be very useful and NAPSR does support that.

Mr. SHARP. In other words, they could call one national number to find out who is the clearinghouse—who is the one-call system in the area they are working? So they are looking for what is the number they should call.

Mr. HURIAUX. Yes, correct, or they could patch them through even to that center. There would be different ways of setting it up and I do think the setting up of such a system should be left to OPS and the one-call industry, and of course the State authorities.

Mr. SHARP. All right. Well, thank you. I am going to recognize my colleague from California because we don't know what the voting situation is going to be here and I want to make sure he has the opportunity to request questions.

Mr. MOORHEAD. Thank you. One thing I was wondering, that 48-hour period that you have, many times there is an emergency plumbing situation in a home where toilets are backed up or sinks are; a 48-hour wait might be a bit much before they got the rotor-rooter or whatever they were going to get. I wonder if this would affect that.

Mr. HURIAUX. Well, Mr. Moorhead, I think obviously allowances have to be made for emergencies. We don't want a pipeline company or a water company having to wait 48 hours while the product is spilling onto the street obviously, but as far as a home situation, I really don't think that anything in the one-call bill would affect the ability of a plumber to make immediate repairs in a home situation.

Mr. MOORHEAD. What type of excavations are covered by the bill? Are gardeners using automatic tillers at risk of becoming criminals if they fail to call the one-call system or do you agree that there should be some willful and knowing standard.

Mr. HURIAUX. I think that is an issue that needs to be addressed in the markup of the bill to make clear how far we are going to go down that road. Clearly if a plumber, to project your first example, if a plumber found they had to dig up the water service line serving the house or the sewer, they ought to be calling the one-call system for location because in that process they could be hitting gas or underground electric lines.

Now, on the other hand, if the basement is flooding, there is a simple solution to that. You turn the water off at the street. You call the water company. So I think these issues can be worked around, but clearly we don't want someone anywhere in the United States using power equipment to excavate a utility line without having locations of conflicting lines which could include gas lines, of course, and electric lines, marked first.

Mr. MOORHEAD. I understand that the Allentown, Pennsylvania accident occurred after a one-call system was used. And the person involved knew that the gas lines were in the vicinity. Should there be greater regulation or participation by facility owners of actual excavation procedures to avoid this type of an accident?

Mr. HURIAUX. I think that is detail that we need to leave to the States depending on the mix of utilities they have to deal with and the density, of course, of the local area. I know in the Washington area, the gas company, for example, and the electric company will often dispatch an inspector to the site of an excavation that has the potential for doing damage, even to their facilities, even after the location of the conflicting facilities have been marked to supervise the matter, to be ready to turn off those facilities if they are inadvertently hit.

Mr. MOORHEAD. Mr. Becker, we didn't intend to leave you out of this. Did you want to make a response?

Mr. BECKER. Thank you. Mr. Moorhead, the issue on locating facility lines, any time you have a high pressure gas line, any type of hazardous underground line, it is our opinion that the owner of the facility should keep a representative on the site until that particular area is cleared of the excavation. Electric lines, high pressure gas lines, hazardous chemical communications, all type of systems like that, you should keep a representative there, not just come out and mark it and leave, and I am not sure if that was your question or not.

Mr. MOORHEAD. Pretty much. It hit a little part of it. I think we have got a vote over on the Floor and I don't want to take all the time.

Mr. SHARP. Do you have any other questions?

Mr. MOORHEAD. Go ahead.

Mr. SHARP. Mr. Pallone, did you have anything?

Mr. PALLONE. I just wanted to say quickly, listening to Mr. Becker's comments about substituting an incentive program for the link to Federal highway funds, one of the reasons why the bill has the link to Federal highway funds instead is because of my concern that we just wouldn't be able to put together an incentive program and have enough money to do it. I think right now there is about $12 million in the escrow in the OPS escrow account.

I don't even know that all that money would be available for grants, but the problem is, to have an effective incentive program you are going to have to come up with some—I would think, big money somewhere, and we don't have it at this point that I am aware of. How much do you estimate it would take if you were to have an incentive for 50 States?

How much money do you think you would need available to make it real?

Mr. BECKER. I don't believe anybody could answer that question.

Mr. PALLONE. Well, that is the—I mean, that is one of the problems really is the lack of funds. Just wanted you to know.

Thank you.

Mr. SHARP. Thank you. Gentlemen, we appreciate your time and attention to this. We have another vote on the House Floor so I think we will bring the hearing to a conclusion and we will welcome continuing comments as we proceed in this toward a markup session.

Thank you.

[Whereupon, at 1:15 p.m., the hearing was adjourned.]

[The following material was received for the record:]

American Public Gas Association
Suite 102
11094-D Lee Highway
Fairfax, VA 22030-5014
703/352-3890

June 23, 1994

The Honorable Philip Sharp
Chairman
House Energy and Commerce
Subcommittee on Energy and Power
331 Ford House Office Building
Washington, DC 20515

Dear Chairman Sharp:

The American Public Gas Association ("APGA")[1] appreciates the opportunity to submit written comments on a national, comprehensive one call notification system and commends the Subcommittee on Energy and Power for moving quickly on this important public policy issue. APGA supports Congressional efforts to enact new legislation requiring notification to and participation with one-call systems and looks forward to working with Congress and the Executive Branch in implementing this legislation.

APGA requests Congress to develop expeditiously a one-call program which is thorough, well-understood, and widely publicized. APGA is committed to supporting a one-call system and has been working with the Department of Transportation's Office of Pipeline Safety over several years to develop guidelines for a one-call system.

The time has come for Congress to enact legislation that ensures all states have laws which:

- establish a one-call system compelling all excavators, including state and local governmental entities, to call before they dig;
- require all underground facilities used for gas, electricity, communication, petroleum, water and sewage to participate so excavators can secure all of the necessary information about a site with one telephone call;
- use a telephone number (such as 344-"DIG") which functions similar to the 911 emergency number. Establishing a single telephone number will make the one-call system easy to use.
- promote public awareness of the benefits of "calling before you dig."

---

[1]APGA is the service organization for this nation's community-owned, not-for-profit public gas utilities that provide retail natural gas service to millions of consumers, businesses, and industries in 34 states. Public gas systems serve approximately 3.8 million of the nation's gas consumers.

- Also, federal legislation should leave it to the states to develop their own sanctions and penalties for those not following the law. Several options for sanctions and penalties to be considered include: civil penalties; administrative fines; judicial injunctive relief; contempt charges (for failure to comply with a judicial or administrative order to comply); strict civil liability; civil liability "multipliers"; excavator disbarments; and publication of lists of violators. Regardless of the method used, each state should have some penalty for not calling or notifying a one-call system.

To maximize the effectiveness and consistency of state and local one-call programs, there should be no exemptions to these standards. In addition, states, local utilities, and excavators should be encouraged to participate by developing a one-call notification program that both works best for them and meets the national standards.

To influence many states, counties, municipalities and excavators to participate in a one-call program, APGA suggest that Congress consider linking federal funding for state and locally sponsored highway, railroad, aviation, and transit projects to participation in a one-call program. Thus, a state, county or municipality would be required to participate in a one-call program in order to receive federal funding for its capital improvement project. Similarly, all excavators could be required to have called a one-call program prior to construction in order to receive payment for work performed under a federal program. APGA encourages active federal oversight to insure that states, counties, municipalities, and excavators are participating in a one-call program.

APGA again thanks the Subcommittee on Energy and Power for holding a hearing on this important issue and urges the Subcommittee to incorporate our comments for one-call notification into legislation being considered. On behalf of public gas systems across the country, APGA looks forward to working with the Subcommittee to develop and implement a safe, effective one-call notification program.

Sincerely,

Robert Cave

Robert S. Cave
Executive Director

ARIZONA BLUE STAKE, INC.
263-1100
1-800-STAKE-IT (OUTSIDE MARICOPA COUNTY)
™ 4747 N. SEVENTH ST., SUITE 440, PHOENIX, AZ 85014

MEMBER ONE CALL SYSTEMS INTERNATIONAL

June 21, 1994

The Honorable Philip R. Sharp
Chairman
Subcommittee on Energy and Power
House of Representatives
H2-331 Ford H.O.B.
Washington, D.C. 20515-6117

Dear Chairman Sharp:

Arizona Blue Stake, Inc. is pleased to submit the enclosed testimony on the "Comprehensive One-Call Notification Act of 1994" in connection with the subcommittee's hearing of June 23rd and the June 24th informal meeting. We would appreciate inclusion of the testimony in the printed record of the hearings.

Thank you for your attention to this matter. If you have any questions or require further information, please do not hesitate to contact me.

Sincerely,

Kim C. Moody

Kim C. Moody
Executive Director

Enclosure

KCM/myk

Mr. Chairman:

Arizona Blue Stake, Inc. owns and operates a one-call notification center in Arizona. We are a private not-for-profit corporation whose 358 owners include public and private utilities, federal, state, county, and municipal governments, and private businesses such as manufacturers, banks, churches, and private citizens. Our business began service to Maricopa County in 1974 and today serves 10 Arizona counties. In 1993 our center received 592,079 notifications from excavators and transmitted 2,265,137 notifications to ourselves, the facility owners, for response. This volume compares to that of the top 5 centers in the United States, and on a per capita basis is the highest notification volume in the United States.

We already support a single nationally accessible 800 number for Arizona excavators with our sister center in Tucson which serves the remaining 5 counties in Arizona. Our center services are identical and almost entirely dictated by Arizona's damage prevention law which requires excavators and facility operators to communicate and respond. Our damage prevention law was created by the collective efforts of all interested parties and subsequently amended, again by collective effort, to include such significant improvements as an enforcement element in 1985 and mandatory participation in a one-call notification center in 1990. Our damage prevention programs, state law and one-call notification center are considered the most comprehensive in the United States and a model in all areas. We have been visited by representatives of foreign countries as well as our colleagues on all levels in the one-call notification center industry. We, by example, take great pride of ownership in our one-call notification center as well as our successful damage prevention programs and state law. We know from our experience that the one-call notification center, which is simply one of many tools in our damage prevention tool box, is not, nor has it ever proven to be, a cause of damage. Comprehensive one-call notification centers will be a product of comprehensive legislation requiring excavators to notify and facility owners to respond.

The owners of Arizona Blue Stake support the concepts of this bill. We are prepared to argue a necessity for this bill and are prepared to demonstrate much has been accomplished and much is in the process of accomplishment with more momentum today than ever within the States. If we are to have a bill, than we would take exception to the language and methodology of sections 4, 6, & 7 which go beyond what is necessary and justifiable to accomplish the goal. We also would bring to the committee's attention the little time, 30-plus days, industry has had to respond to this measure and benefit the composers of this bill with our 30 years of knowledge and experience. This bill comes at the height of the excavation season in the United States leaving all interested parties overwhelmed with activity and less likely to give sufficient and expedient response.

We calculate there are 45 states with damage prevention laws. Forty-eight states have a one-call notification center. Of the 48 states, 35 are single statewide centers. There are multiple centers in 13 states. Of those 13, 2 are Arizona and California, and have created single 800 numbers to easily direct the caller to the correct center. One-call notification center programs and damage prevention laws are actively being discussed and developed in the 5 states remaining without a law and the 2 states that do not currently have a center. Of the existing states with damage prevention laws many are in the process of improving and making their laws more comprehensive. There is incredible momentum in this area that goes unrecognized in this bill. This momentum can be increased and supported by our government in alternative ways other than a one size fits all solution for the States. Population demographics, geography, and seasonal weather patterns play major roles in a state's program requirements.

Basic element of a successful damage prevention program are:

- Require excavators to notify facility operators and ascertain the identification and location of underground facilities before they dig.

- Require all facility operators to respond to all excavator notifications.

- Require facility operators to establish and participate in a one-call notification center.

- Provide for penalties and enforcement for violating the requirements.

Section 4 outlines a nationwide 800 number to be established by the Secretary of Transportation. Although the modified language in the June 16th discussion draft gives greater input to industry to help establish this nationwide number, the language still implies that all calls to a one-call notification center will be routed through the nationwide number. Industry has already adopted a project that would provide a national 800 number which would route unknowing callers to the correct one-call notification center. The industry's program is designed to be a directory assistance or information and referral program accomplishing this goal in a much more practical and inexpensive manner. Industry should be given the opportunity to design and implement its own services. If there is doubt industry will accomplish this goal, then we suggest language that would require industry to do this within one year and if not accomplished would require the Secretary of Transportation to establish the service. At a minimum, the language should outline the program as intended, as we are told, an information and referral service. We feel the nationwide 800 number will be more successful if the responsibility remains in the control of the one-call notification center experts. Also routing calls within a single entity, such as Pizza Hut, is a different practice than routing calls to different companies. As we have learned through Arizona's experience of establishing the first call routing program between two separate companies, long distance carriers are not in a position to break down billing between the two entities. Therefore, identifying the costs to each center requires a joint effort by both centers. Also with this issue, this proposal comes before industry has had the opportunity to accomplish this service for ourselves and it must be noted our efforts began before this bill was introduced and we feel the government should welcome our industry's willingness to be self-sufficient.

Section 5 provides for $12,000,000 dollars in grant funding to establish programs. First, we are not sure this money can be used for this purpose and secondly, industry simply doesn't need it. Financial resources are already available to non-participants of a one-call notification center and/or damage prevention programs. USWEST Communications reported a 60% reduction in damages its first year of participation which more than paid for their participation in the program and center. Unfortunately, other than these first year examples, our industry will never be able to justify the incredible savings in dollars and lives because we participated as facility operators and excavators in a damage prevention program and one-call notification center. If we could justify it somehow, this argument of funding would be a dead issue. It is simply a matter of turning the negative dollars currently being spent into positive dollars and in the process the participant <u>will</u> experience a savings. Especially municipal, county, and state governments, which can utilize the excavators notice to check permit requirements. One Arizona city has not only covered entirely their cost of participating in the one-call notification center but has recovered permitting fees they would not have, otherwise. Pittsburgh, Pennsylvania is another success story in this regard.

Section 6, provision 1, by its inclusion, places regulation by the state over our business's daily operation. With regard to section 7; since Arizona has an active enforcement program, the state, in this context, refers to the Arizona Corporation Commission. This interpretation of regulatory jurisdiction has already been conveyed to us by representatives of the Commission. We know of no evidence or justification that our business should be regulated. To the contrary, statements in section 1 of this bill point to how effective one-call notification centers are. We must bring to your attention the New Jersey incident which has prompted this bill was not caused by the one-call notification center. The inclusion of such operational detail is not typical of a model state damage prevention law. Arizona law only addresses the one-call notification center in its definition section and only refers to it again in its requirement that facility owners participate in the one-call notification center. In fact, most are surprised to learn Arizona law does not require the excavator to notify the one-call notification center. So how has our program so successfully motivated the excavator to participate? Service, the only reason the center exists.

We also know that at least 65% of excavation notices we receive are generated by the owners of the center by their excavation personnel. Other states report similar if not higher percentages. Therefore, the center's customers are also its owners and a center has no motivation to provide substandard service at their owner's expense; which addresses provision number 9 of section 6 which proposes the state establish fee schedules for owners instead of the owners establishing their own fee schedules for themselves. What business would force their overhead costs to increase without realizing a profit? Owners are not motivated to do this nor do I believe there is any evidence that center owners do this.

Section 7 provides for penalties and an enforcement program. We believe the suggested penalties amounts are excessive without merit and will result in excavators afraid to report damage. Expenses associated with damage to underground facilities are a significant deterrent today for any excavator and facility owner who wants to stay in business. Secondly, adding the element of enforcement will generate more savings than the penalties proposed in the bill. If penalties are set too high, excavators will be reluctant to report minor damage to a facility operator; the same kind of damage that, if covered up and left unrepaired, could potentially be disastrous. Arizona has an enforcement program unlike any other state's with penalties of $2000 per violation and includes injunctive relief. This amount is very similar to other state's damage prevention law penalties. Our enforcement program usually waives or reduces fines if the violator submits to education; it has been effective and has produced long term results with few, if any, second time offenders. High penalties outlined in the Pipeline Reauthorization Act of 1986 have had little to no impact on the Arizona program. What has is enforcement through education, cost of damages, and the ability to continue or obtain insurance coverage.

We need to direct efforts towards substandard programs without penalizing a model program, like Arizona Blue Stake. This bill subjects Arizona to adopt a substandard program, request the State to submit a petition for waiver to recognize our damage prevention program, state law, and center exceed the requirements or see our Arizona be penalized over 2 million dollars in much needed transportation funding. The Arizona community doesn't deserve this reward for our successful, self-sustaining program and center.

For your consideration Arizona Blue Stake has developed language for this bill which is attached. We are prepared to work with the subcommittee to effect these important changes and to assist the subcommittee in any other appropriate way.

Thank you for your efforts and for the opportunity to submit testimony.

attachment

Comments to H.R. 4394 Discussion Draft of June 16, 1994

SEC. 3. DEFINITIONS.

(2) "excavation" means any operation in which earth, rock, or other material in the ground is moved, removed, or otherwise displaced by means of any tools, equipment, or any explosive. Strike rest of sentence.

Comment: The current language excludes hand tool excavation. Most excavator injuries are the result of hand excavation. No person or industry should be exempt from this program. Alternate notification programs should be provided to address agriculture, mining, and continuous road grading operations. I believe the language that provides this is section 9.

(4) "facility operator" means any person who owns or operates an underground facility, except for any person who is the owner of real property wherein are located underground facilities for the purpose of furnishing services or materials only to himself.

Comments: Although we have seen this language go from one extreme to the other, we believe all industry would agree the intent is not to make a homeowner a facility operator and participate in the one-call notification program. Although any computer system can be developed to in Arizona's case, manage 3 million plus homeowners, it would be extremely expensive. This suggested language would not exempt owners of tenant properties who should be responsible for the safety of their tenants and excavators required to work around owner's underground facilities. Many apartment complexes, office and industrial complexes, trailer parks, etc., own master metered gas, electric, and communications systems. It is a landlord's obligation to operate these systems safely.

(6) "pipeline facilities" (Strike all)

Comments: We do not understand why this definition has been added? This language will exclude gathering facilities, that carry products that have not, in most cases, been processed for delivery and facilities transporting hazardous liquids at twenty percent or less of the specified minimum yield strength of the pipeline would not be required to participate.

(9) "State program" means the legislative requirements of a State to comply with the provisions of this Act. Strike rest of sentence.

Comments: By changing the word "program" to "legislative requirements" the intent of the Act to provide comprehensive programs will be achieved and we believe with greater success. This language would no longer require States to take over established center business yet does not preclude the State from doing so, if necessary.

84-088 O - 94 - 7

(11) "one-call notification center" means an organization of owners or operators of underground facilities which provides a telephone notification service for the purpose of receiving and distributing to its members advance notifications from persons regarding excavations.

Comments: This definition takes the place of the section 6(1A-G). See comments in section 6.

SEC. 4. NATIONWIDE NUMBER PROGRAM.

Within one year after the date of enactment of this Act, all state one-call notification centers shall establish a nationwide telephone number excavators can use in contacting the correct one-call notification center. One-call notification centers must fund the nationwide number program. If this requirement has not been fulfilled by the expiration of the time allotment, the Secretary shall establish the nationwide number program.

Comments: The program that was proposed would require all calls to one-call notification centers in the United States to filter through the nationwide number. This method would be extremely technically challenging, very expensive and would decrease efficiency. We believe the intent is to provide a safety net so that an unknowing excavator could contact a nationwide number and be routed or referred to the correct center. A nationwide number and this approach are not new concepts. OCSI just adopted this concept as a service goal as well as interstate facility operators have been lobbying such a program for some time. The system envisioned would operate like a directory assistance service with a call-routing feature. The suggested language would provide that opportunity and makes provisions if efforts were unsuccessful.

SEC. 5. STATE PROGRAM.

(a) REQUIREMENT.-Each State shall, within 3 years after the date of enactment of this Act, provide for the establishment of comprehensive statewide damage prevention legislation (Strike including a one-call notification system) in accordance with this Act, to protect all underground facilities from damage due to any excavation.

Comments: Industry has defined these programs as a damage prevention system. This was a deliberate effort by industry to communicate awareness that requirements to prevent damage go way beyond making one call. The distinction is important and would make this legislation easier and less confusing to understand. Including a one-call notification system was removed since it is addressed in Section 6.

(b) Strike all

Comments: As indicated in our testimony, where is the money coming from and can it be used for this purpose? Funding is already available for industry. This appropriation is not necessary.

SEC. 6. ELEMENTS OF STATE PROGRAM.

Each state program under section 5(a) shall, at a minimum--

(1) provide for facility operators to establish a one-call notification center which shall-

(A) operate in all areas of the State and not duplicate the geographical coverage of other systems;
(B) provide for 24-hour coverage for emergency response;
(C) provide for a fair and appropriate schedule of fees to cover the costs of establishing, maintaining, and operating the one-call notification center;
(D) inform excavators of the identity of facility operators that will be notified and what types of facilities they may identify;
(E) Inform excavators of any procedures that the State has determined must be followed when excavating;

Comments: We are suggesting a re-write of this section. This language requires States to enact legislation that provides for the establishment of one-call notification centers by facility operators and preserve the business ownership of existing centers as well as insure program progression. I have removed many of the items in section 6(1) as I am not aware of a center that does not already perform these requirements. I left the ones that are needed to bring about consistency in the center programs. Also the definition of a one-call notification center has been added to Section 3 that covers the removed requirements. Lastly, by mandating legislation as verses state control, facility operators will decide what fees and services are necessary to service themselves and the community. With the inclusion of language that requires facility operators to establish fees for themselves in the legislation, facility operators would be able to protest their fees and services to the enforcement agency if necessary. This language does not preclude a State from establishing a one-call notification center and preserves the brain trust necessary for a successful center program.

(4) Moved to section 6, part 1(E)

(7) Provide for procedures to improve the records of underground facilities, including adding to appropriate record systems any information about previously unknown underground facility locations when such information is discovered;

Comment: Arizona is the only State I am aware of that has a records requirement. The terms mapping, as-built, etc. were debated at length in our legislature. It was decided the word "records" was the most accurate terminology because of alternative and even more efficient methods of identification other than mapping.

(9) Moved to section 6, part 1(C)

Add (11) require all facility operators to participate in a one-call notification center;

Comments: This requirement appears to be implied and should be addressed directly.

Add (12) require that no excavator shall begin excavation, regardless of facility operator requirements in Sec. 6(4) until all underground facilities have been identified or facility operator has notified excavator that identification is unnecessary.

Comments: This is an important safety check system. As humanly intense as this program is, this requires the excavator to have all facilities identified prior to excavation and provides a check system if something has gone wrong. Arizona law provides that the excavator can charge the facility operator for downtime when the operator has not complied within the response time. Further, enforcement will address facility operators not in compliance. Excavators assume if nothing is marked then the facility operator does not have facilities in the area. This is a dangerous assumption. By emphasizing this in the State requirements there is no room for assumption.

SECTION 7. PENALTIES AND ENFORCEMENT.

Comments: As indicated in testimony we believe a penalty should not exceed $2,000 per violation of this Act.

THE ASSOCIATED GENERAL CONTRACTORS OF AMERICA
1957 E Street, N.W. • Washington, D.C. 20006 • (202) 393-2040 • FAX (202) 347-4004

July 29, 1994

The Honorable Philip R. Sharp
Chairman
Subcommittee on Energy and Power
U.S. House of Representatives
Washington, D.C. 20515-6117

Dear Mr. Sharp:

Thank you for providing AGC with the opportunity to comment on the amendment in the nature of a substitute to H.R. 4394, "The Comprehensive One-Call Notification Act of 1994". The following are offered in response to the questions posed in your July 11, 1994 letter:

1) AGC believes that universal, mandatory participation is the minimum requirement for having successful One-Call systems in operation around the country. From the excavator's point of view, it is important to know that all existing underground facilities are located before excavation activities begin. The best way to ensure this is to require that all facility owners participate in the system. Confidence in One-Call systems is undermined when owners do not respond in a timely fashion or when there is uncertainty about which owners will be coming to the site of the construction to mark facilities. This can be avoided by having universal participation in One-Call systems and requiring the One-Call system to contact the excavator to indicate which owners in the excavation area have facilities which need to be marked.

AGC believes that the primary factor undermining the success of existing One-Call systems is that certain underground facilities are exempted from coverage. For this reason AGC is reluctant to support any exemptions from participation. However, it is AGC's understanding that gathering lines and flow lines are generally located on property owned or controlled by the same company which owns the lines. Furthermore, it is AGC's understanding that the only excavation activities which are likely to occur on these lands would be undertaken by or controlled by the owner of the lines. Under these limited circumstances AGC can support an exemption for gathering lines and flow lines. Flow lines and gathering lines which cross rights-of-way would still have to be marked by the owner.

2) AGC believes that states should be relied on to develop their own One-Call systems taking into consideration their own unique circumstances. However, AGC believes that the Federal legislation can prompt the development of good One-Call systems by: requiring mandatory, universal participation; providing funds to assist states in setting up their systems; providing models for states to use in fashioning their systems; and, developing educational programs to increase the understanding and use of the systems.

AGC believes that an important part of this whole process, which has been overlooked, are the engineers who design new projects to be constructed. When new construction projects are in the design phase, it is an ideal time to ensure that underground facilities are accurately located. In this way, the location of utilities can be included in the plans and specifications before any excavation begins. Engineers should design new projects in such a way that they avoid or at least minimize disruption of existing underground facilities. AGC recommends that this concept be included in a model national plan.

3) AGC is pleased that the substitute version of H.R. 4394 does not specify penalty amounts for violations of state requirements. While AGC realizes that penalties may be necessary to enforce the requirements of the state programs, AGC believes that One-Call programs should not rely too heavily on penalties. AGC believes that education and cooperation are the more productive means for achieving the goal. Penalties only create an adversarial environment which is counterproductive. AGC also believes that penalties are often used by governmental bodies as revenue raising sources rather than for their intended purposes. When this happens the cooperative nature of these programs is severely undermined.

4) AGC believes that public education is the most important aspect of One-Call notification systems. In addition to public service announcements, AGC recommends the following: programs at excavators events such as conventions or monthly meetings, video tapes which could be shown to excavator crews at safety meetings, programs as part of excavator training sessions. For example, AGC has 101 chapters around the country which have monthly meetings, annual conventions and sponsor many different types of training programs. These should be utilized. It is likely that other excavator organizations, such as utility companies and public works departments also have a local network which could be utilized for this purpose.

Thank you for your interest in addressing this problem of national concern. AGC looks forward to working with the Congress and the Department of Transportation in developing a program which can produce positive results.

Sincerely,

Brian Deery
Director
Municipal-Utilities Division

Edwin L. Harper
President and
Chief Executive Officer

ASSOCIATION
OF AMERICAN
RAILROADS

July 14, 1994

The Honorable Philip R. Sharp
Chairman
Subcommittee on Energy and Power
Committee on Energy and Commerce
U.S. House of Representatives
Washington, D.C. 20515

Re: **H.R. 4394 - Comprehensive One-Call Notification Act of 1994**

Dear Mr. Chairman:

The Association of American Railroads (AAR), a trade association representing freight railroads and Amtrak, respectfully requests that the following comments on H.R. 4394, the Comprehensive One-Call Notification Act of 1994, be made part of the record for the hearing your Subcommittee held on June 23. While AAR supports legislative efforts to prevent accidents involving underground facilities, AAR believes this bill would impose unnecessary burdens on the railroad industry.

AAR's primary criticism of H.R. 4394 concerns the definition of "excavation." Section 3 of the bill defines excavation as "any operation in which earth, rock, or other material in the ground is moved, removed, or otherwise displaced by means of any mechanized tools or equipment, or any explosive, but shall not include the tilling of soil for agricultural purposes to a depth of 18 inches or less; . . ." This definition would include regular track and trackbed maintenance activities performed every day by the railroad industry.

Under H.R. 4394, ballast cleaning, general ballast work, ditch cleaning and the replacement of ties would be considered excavation. However, these maintenance activities disturb the soil in only an incidental manner. These activities must be performed on a regular, and in some cases daily, basis by railroads to ensure the safe and efficient operation of their trains. H.R. 4394 would require railroads to contact the one-call notification system at least two business days before beginning any of

50 F Street, N.W., Washington, D.C. 20001 (202) 639-2402

these activities on a section of track. Such a requirement would adversely affect the efficient operation of the railroad industry.

Regular track and trackbed maintenance do not damage underground facilities. Railroads are almost always aware of the location of underground facilities on their rights-of-way and can perform necessary maintenance activities without threatening such facilities. Should there be any question about the location of an underground facility, the industry has a standard practice of contacting the relevant one-call system in order to ensure that performance of the contemplated maintenance activity is safe.

In consideration of these unique circumstances, an exemption from the definition of "excavation" for regular track and trackbed maintenance activities would be appropriate. H.R. 4394 already excludes "the tilling of soil for agricultural purposes to a depth of 18 inches or less" from the definition of excavation. Just as there is justification for excluding agricultural tilling to a depth of one and one-half feet, there is at least as strong a justification for excluding rail track and trackbed maintenance activities.

AAR urges the Subcommittee to provide an exemption in H.R. 4394 for regular track and trackbed maintenance activities. Specifically, AAR suggests that section 3 of the bill be revised to exclude from the definition of excavation "the movement, removal or displacement of soil for the purpose of regular track and trackbed maintenance activities by a railroad." Such language would ensure that railroads can continue to operate efficiently without imposing any threat to public safety.

AAR has two additional concerns about the present bill. First, AAR urges the Subcommittee to add language to H.R. 4394 stating that the provisions of the bill do not override any contractual obligation which requires an excavator to notify a railroad prior to excavating. Railroads regularly enter into agreements requiring an excavator on a railroad's right-of-way to call that railroad before excavating. Such contractual obligations should not be superseded by this legislation.

Finally, the definition of "emergency circumstances" at section 6, subsection (8) of the bill

The Honorable Philip R. Sharp
July 14, 1994
Page 3

should be broadened to include washouts and other natural disasters, which can block train operations and cause delays in moving mainline traffic. Under H.R. 4394 as presently drafted, a washout on a Friday night would necessitate waiting two business days (until the following Wednesday) before excavation could begin and traffic could move again. This situation is unacceptable. If washouts and other natural disasters were included in the definition of "emergency circumstances," however, railroads would receive an appropriate waiver of timely compliance with the one-call notification system and could begin necessary excavation work immediately.

Sincerely,

Attachment

ASSOCIATION OF AMERICAN RAILROADS

PROPOSED REVISIONS TO H.R. 4394

1. Subsection 3 (2) - Add the following language at the end of this subsection, between the word "less" and the ";": "or the movement, removal or displacement of soil for the purpose of regular track and trackbed maintenance activities by a railroad"

2. Section 6 - Add the following language as a new paragraph at the end of this section, after subsection 6 (8): "Nothing in any State program shall supersede a contractual obligation which requires an excavator to notify a railroad prior to performing an excavation."

3. Subsection 6 (8) - Add the following language in the middle of this subsection, after the word "safety,": "or in circumstances in which a washout or other natural disaster occurs along railroad track,"

U.S. Department
of Transportation

**Research and
Special Programs
Administration**

The Administrator

400 Seventh Street, S.W.
Washington, D.C. 20590

AUG 18 1994

The Honorable Philip R. Sharp
Chairman, Subcommittee on Energy and Power
Committee on Energy and Power
U.S. House of Representatives
Washington, DC 20515

Dear Mr. Chairman:

The Research and Special Programs Administration (RSPA) appreciated the opportunity to appear before the Subcommittee on Energy and Power on June 23, 1994, to testify on H.R. 4394, the "Comprehensive One Call Notification Act of 1994" and on H.R. 6414, the "Natural Gas Pipeline Safety Improvement Act of 1994."

Our answers to the follow-up questions that were included in your letter of July 8, 1994, are enclosed for inclusion in the hearing record. If we can be of further assistance in this matter, please contact me or Ms. Patricia Klinger, who handles our congressional inquiries, at (202) 366-4831.

Sincerely,

Ana Sol Gutiérrez
Acting Administrator

Enclosure

QUESTION 1:

In your testimony, you oppose the suggested funding mechanism that provides grants to the States for improvement of their one-call systems. Do you have an alternative proposal for providing incentive money to promote effective one-call systems?

ANSWER: The $17 million in the user fee fund is the unappropriated funds initially set up when the user fee was initiated. This money is presently used for cash flow between the beginning of the fiscal year and the receipt of fees from pipeline operators. Taking an immediate $12 million from the fund could cause a cash flow problem.

An alternative would be to provide the $12 million over 3 years ($4 million per year) from the user fee "float" of $17 million.

QUESTION 2:

I understand that you object to the use of Federal Highway Funds as a sanction against States who do not establish a One-Call Program that is at least as effective as the Act would require. Can you suggest an alternative mechanism that would ensure that States follow the directives set forth in this Act?

ANSWER: The sanction of cutting Federal Highway Funds has been removed in the July 14, 1994, version of the bill. We believe the new proposal in Section 7 which would require the Secretary to "review, and report to Congress on, the extent to which any policies, programs, and procedures of the Department of Transportation could be used to achieve the purposes of the Act" would be effective for identifying within the Department of Transportation methods for increasing participation in one-call damage prevention programs.

QUESTION 3:

In your opinion, how much money is necessary to implement a grant program for improving one-call systems that will actually have some impact on the States? How much do one-call systems cost? How much do they save in terms of damage prevention?

ANSWER: In its amended Fiscal Year 1995 budget request, OPS has asked for $1 million for improving one-call systems .

Since all states except Hawaii, North Dakota and most of Alaska, already have one-call service, only these three would need to have start-up costs considered. This cost is estimated at $100,000 to $250,000 per state for employees, hardware and telecommunications. There is a cost associated with establishing the database of operators. Informal contacts with representatives of the one-call community indicate that costs normally run about $10,000 per county. Annual operating costs are usually covered by fees paid by facility operators and are based on number of calls they receive from the one-call center, by the area served, or by a combination of these.

While there is no nationwide data on the damage prevented through participation in one-call, there is some state-specific data available. US West in Arizona saw a 60% decrease in damage in the first 18 months after the one-call center opened. Northern States Power Company in Minnesota reported that the total number of pipeline dig-ins grew by only about 2 percent while digging activity increased by 71 percent in the same 2 1/2-year time interval. Nationwide, AT&T showed a 90 percent reduction in hits in the 18 months after January 1992, when AT&T assigned 53 full time "Contact Representatives" to work with excavators to better utilize the one-call system. By using a strong education program to get excavators and facility operators to work together, Alberta, Canada, has had a reduction from 24.11 damages per 1,000 locates in 1983 to 5.80 in 1993.

QUESTION 4:

Would you support the collection of user fees from all underground facility operators to help finance the one-call incentive program?

ANSWER: Yes. Most one-call centers are funded by the member utility operators. Many, if not most, of these operators would be willing to spend a bit more to improve the system. This would be the most efficient means of applying a user fee in contrast to establishing a new fee structure to be paid directly to a state. One means of incentive funding would be to earmark a certain amount per-call or per-grid to be used for such purposes as education, facility improvement, and enforcement.

QUESTION 5:

Do you think the penalty provisions in H.R. 4394 are too high? What type of penalty structure should be included in Federal legislation?

ANSWER: The specific penalty provisions for violations of the one-call requirements in the July 14, 1994 version of the bill have been replaced by provisions allowing for state discretion in providing for administrative or civil penalties. We support this approach, as well as the provisions for increased penalties for violations which result in serious accidents, and higher penalties when damage results from a violation of the damage prevention laws.

QUESTION 6:

What are the essential elements that should be included in Federal legislation as minimum requirements of a one-call system? (Please refer to my discussion draft of the amendment in the nature of a substitute to H.R. 4394.)

ANSWER: We believe that the following elements, as articulated in HR 4394, are essential for an effective one-call system:

- Mandatory participation by all underground facility operators.
- Only limited and narrowly drawn exceptions to the requirement that all excavators must call before they dig.
- Administrative penalties and enforcement that is simple to administer.
- A public education campaign to teach people to call before they dig, and then how to dig safely around underground facilities.
- Twenty-four hour per day, seven day a week service at one-call centers to handle requests for locates on emergency basis.
- Statewide coverage by a one-call center without overlapping coverage.

QUESTION 7:

Should there be any exceptions to universal, mandatory participation in a one-call system? Do you think that exempting gathering lines and flow lines undermines our goal of universal, mandatory participation? Are there special circumstances or situations that would justify the exemption of such lines?

ANSWER: Any exceptions deemed necessary should be narrowly drawn and should be evaluated on whether the exception would significantly increase the probability of damage to underground facilities that would result in economic damage or injury to people, property or the environment. The July 14 version of the bill is sufficiently narrow in its exemption that we feel it does not undermine the goals of the bill. The gathering line industry has expressed the need to exempt those lines that are under their direct control, i.e., on their controlled access property where no one else can access or dig. We can support this type of exemption, because as the sole users of the property, the only operator notified would be themselves.

QUESTION 8:

Would it be helpful to the Secretary to involve the FCC in the establishment of the nationwide phone numbering system or are you confident that the Department, in consultation with underground facility operators, excavators, and one-call system representatives, will be able determine the most effective use of such a number?

ANSWER: We do not believe that we would need to involve the FCC because the type of phone number the bill apparently addresses does not require the involvement of the FCC. We do not interpret the intent of the bill to require all calls, nationwide, to go through the national number. It is our understanding that only those calls where an excavator does not know the correct number, primarily if he would be excavating out of state, would use the nationwide number. The national number should not preclude a statewide number but should be used in conjunction with a statewide number.

QUESTION 9:

How important is the role of public education in the use of one-call systems? Do you have any suggestions, other than the use of public service announcements which has been provided for in the bill, on how we can best educate the public?

ANSWER: Public education is of the utmost importance in the success of underground facility damage prevention. Public education must go beyond simply advertising a one-call number to extensive and recurring education in safe digging techniques, since marked facilities are hit more often than unmarked facilities. One successful approach has been the involvement of local Underground Coordinating Committees which meet with excavator groups to discuss safe dig practices. These committees meet regularly with excavators and can build on these already existing relationships. Many unions and contractor's organizations like to participate in these programs. RSPA will also seek to extend damage prevention education into the schools where the message can be communicated early and the children become messengers to their parents.

QUESTION 10:

Do you agree or disagree with Chair Moler's assessment of siting provisions referred to in H.R. 4616? Do you think that FERC should have a greater role in considering safety when siting pipelines?

ANSWER: We are working with the committee to craft a substitute provision to address siting. We do believe that siting as a function of risk must be a consideration when constructing a pipeline.

Paul E. Devaney
Member of Technical Staff
Room MRE-2D177
445 South Street
Morristown, NJ 07960
201-829-3272
FAX 201-829-2622

8/10/94
6/23/94
8/10/94

July 26, 1994

Chairman Philip R. Sharp
U. S. House of Representatives
Committee on Energy and Commerce
Subcommittee on Energy and Power
Room H2-331
Ford House Office Building
Washington, DC 20515-6117

Chairman Sharp

Thank you for the opportunity to be involved in the very important legislation effort of H.R. 4394, the "Comprehensive One-Call Notification Act of 1994."

I have received the follow-up question stemming from the June 23, 1994 hearing and have provided the answers to those question as an attachment to this letter, using the July 14, 1994 Substitute amendment to H.R. 4394, as the basis for my response.

I must first qualify my position as it relates to the Network Reliability Council - Fiber Cable Focus Group, for which my original testimony was based. The Council established in 1992 by the Federal Communications Commission has recently undergone revisions in membership and modifications to its charter. The Council had also established a Network Reliability Steering Committee (NRSC), under the auspices of the Alliance for Telecommunications Industry Solutions, to monitor network reliability. The original Fiber Cable Focus Group of which I was a former member no longer exists to review your questions. I have however, attempted to answer the questions based upon the findings and conclusions of the original Fiber Cable Focus Group.

Again, I thank you for this opportunity and should you need additional information relating to this legislation please contact me.

Respectfully,

Paul E. Devaney

Attachment Page 1 of 2

1. **The Suggestion has been made to restructure this legislation towards a more incentive - based approach, rather than a mandated approach. Would you support this concept?**

The question as to the approach of legislation was not a topic discussed by the NRC - Focus Group. The Group did however recognize the necessity of continuing to search for innovative approaches to damage prevention.

Would you support the collection of user fees from all underground facility operators to help finance a one-call incentive program?

Underground facility owners already finance the application of call-before-you-dig operations. This group could not make commitments for individual facility owners nor did it make recommendations as it concerned user fees.

2. **What is the most effective means to utilize a single toll-free nationwide telephone number system?**

The use of a toll-free nationwide number system was reviewed by the group. No significant information was identified that the lack of this type of number was a route cause of cable damage. The group did identify that no notification accounted for a large fraction of dig-ups. Ensuring that excavators provide consistent, accurate, and prior notification of excavation activities is a principal line of defense for preventing cable dig-ups. The use of a single number could help to provide a consistent and recognizable instrument.

Should the FCC be given a role in helping to determine the most effective use of such a number?

While the Federal Communication Commission could be available for consultation, the North American Numbering Plan (NANP) administered by Bellcore as well as industry committees could be the source of information and expertise in establishing this number.

3. **What are the essential elements that should be included in Federal legislation as minimum requirements of a one-call system?**

Listed in the Compendium of Technical Papers from the Network Reliability: A Report to the Nation, under Fiber Optic Cable Dig-Ups: "Causes and Cures", Section A, Appendix 7 - Pages 71 to 76 provided Model Legislation from the Group. I have provided a copy of these pages with this attachment.

Attachment Page 2 of 2

4. **Should there be any exceptions to universal, mandatory participation in a one-call system?**

No exceptions should be allowed.

5. **Do you think the penalty provisions in H.R. 4394 are too high?**

No

The discussion draft dated July 14, 1994 lack in increased penalties for hitting a fiber cable system. A change to line 22 after $50,000, <u>effecting 30,000 or more citizens for more than 30 minutes,</u> would help in stopping fiber cable cuts.

What type of penalty structure should be included in Federal legislation?

See Section 5. Penalties - page 75 of attached pages.

6. **How important is the role of public education in the use of one-call systems?**

One of the most important aspects of damage prevention will be public education, on how the system works and the dangers associated with not requiring the use of notification by contractors, municipalities, utilities, state agencies and them selves.

Do you have any suggestions, other than the use of public service announcements which has been provided for in the bill, on how we can best communicate the importance of this issue to the public?

The Group discussed the use of placing notification and education information in phone directories, possibly a warning or statement of laws under the EXCAVATING section.

It is imperative that facility owners strive to educate excavators, contractors, and the public at large about call-before-you-dig procedures and one-call legislation, to shift the emphasis from damage repair to damage prevention.

A bill to be entitled

An act relating to underground facilities, creating the Underground Facility Damage Prevention and Safety Act; providing legislative intent; providing definitions; providing for participation by underground facility operators; providing procedures; requiring notice of excavations and demolitions; providing for liability of certain entities and the system providing civil and criminal penalties; providing an exemption for single family residential owner; providing an exception for emergency excavations under certain circumstances; providing for applicability and construction; providing severability; providing an effective date.

Be It Enacted by the Legislature of State(s) (TBD)

Section 1. Short title; legislative intent.--

(1) This act may be cited as the "Underground Facility Damage Prevention and Safety Act."

(2) It is the purpose of this act to: aid the public by preventing injury to persons or property and the interruption of services resulting from damage to an underground facility caused by excavation or demolition operations.

Section 2. Definitions.-- As used in this act:

(1) "Business hours" means all hours of a day.

(2) "Damage" means any impact upon or contact with, including, without limitation, penetrating, striking, scraping, displacing, or denting, however slight, the protective coating, housing, or other protective devices of any underground facility, or the removal or weakening of any lateral or vertical support from any underground facility, or the severance, partial or complete, of any underground facility.

(3) "Demolish" or "demolition" means any operation by which a structure or mass of material is wrecked, razed, rended, moved, or removed by means of any tool, equipment, or discharge of explosives, or any disturbances of the earth in any manner on public or private lands which could damage any underground facility.

(4) "Excavate" or "excavation" means any manmade cut, cavity, trench, or depression in the earth's surface, formed by removal of earth, intended to change the grade or level of land, or to penetrate or disturb the surface of the earth, but does not include tilling of soil for agricultural purposes.

(5) "Excavator" or "excavating contractor" means any person performing excavation or demolition operations.

(6) "Operator" means any person who furnishes or transports materials or services by means of an underground facility.

(7) "Person" means any individual, firm, joint venture, partnership, corporation, association, municipality or other political subdivision, governmental unit, department, or agency, and includes any trustee, receiver, assignee, or personal representative of a person.

(8) "Underground facility" means any public or private personal property which is buried, placed below ground, or submerged on any operator's right-of-way, easement, or permitted use which is being used or will be used in connection with the storage or conveyance of water, sewage, electronic, telephonic, or telegraphic communications, electric energy, oil, petroleum products, natural gas, optical signals, or other substances, and includes, but is not limited to, pipes, sewers, conduits, cables, valves, and lines. For purposes of this act, a liquified petroleum gas line is not an underground facility unless such line is subject to the requirements of Title 49, Code of Federal Regulations, provided there is no encroachment on any operator's right-of-way, easement, or permitted used.

(9) "System" means a telephone notification system as provided in this act.

Section 3a. Procedures.--

(1)(a) Not less than 2 nor more than 5 business days before beginning any excavation or demolition, an excavator shall provide the following information to each operator:

1. The name of the individual who provided notification and the name, address, including the street address, city, state, zip code, and telephone number of his employer.

2. The name and telephone of the representative for the excavator.

3. The county, the city or closest city and the street address of the closest street, road, or intersection to the location where the excavation or demolition is to be performed, and the construction limits of the excavation or demolition.

4. The commencement date and anticipated duration of the excavation or demolition.

5. Whether machinery will be used for the excavation or demolition.

6. The person or entity for whom the work is to be done.

7. The type of work to be done.

8. The approximate depth of the excavation.

(1)(b) The excavator shall provide such information by calling the operator's toll-free number.

1. Each notification shall be recorded to document compliance with the act. Such record may be made by means of electronic, mechanical, or any other method of all incoming and outgoing wire and oral communications concerning location requests. Such records shall be kept for a period of 5 years and, upon written request, shall be available to the excavator making the request.

2. If an operator determines that a proposed excavation or demolition is in proximity to or in conflict with an underground facility of the operator, the operator shall identify the horizontal route to within 24 inches form the outer edge of either side of the underground facility by the use of stakes, paint, flags, or other suitable means within 48 hours. After the time the notification is received, if the operator is unable to respond within such time, the operator shall communicate with the person making the request and negotiate a new schedule and time that is agreeable to, and should not unreasonably delay, the excavator.

3a. An excavator shall avoid excavation it the area described in the notice given pursuant to subsection (1) until each operator underground facility has been marked and located, or for 48 hours, after notification, whichever occurs first. If an operator has not located and marked its underground facilities within 48 hours, after notification under the system the excavator may proceed with the excavation, provided the excavator does so with reasonable care, and provided, further, that detection equipment or other acceptable means to locate underground facilities are used.

3b. An excavator shall not excavate or demolish in the area described in the notice given pursuant to subsection (1) until all operator underground facilities have been marked and located, or removed.

4. An operator which certifies that it does not have accurate information concerning the exact location of its underground facilities is exempt from the requirements of subsection 3, but shall provide the best available information to the excavator in order to comply with the requirements of this section. An excavator is not liable for any damage to an underground facility under the exemption in this subsection if the excavation or demolition is performed with reasonable care and detection equipment or other acceptable means to locate underground facilities is used.

5. If extraordinary circumstances exist, an operator shall notify the excavator of the operator's

A

inability to comply with this section. For the proposes of this section, "extraordinary circumstances" means circumstances other than normal operating conditions which exist and make it impractical for an operator to comply with the provisions of the act. The operator is relieved of responsibility for compliance under the law during the periods that the extraordinary circumstances exist and shall properly notify the excavator when the extraordinary circumstances cease to exist.

(6) If an operator determines that the excavation or demolition is not near an existing underground facility of the operator, the operator shall notify the excavator within 48 hours, excluding days other than business days, from the time of the notification that no conflict exists and that the excavation or demolition area is clear. An excavator is responsible for contacting the operator if a facility is not marked.

(7) An operator shall use the recommended guidelines for uniform temporary marking of underground facilities as approved by the Utility Location and Coordinating Council of the American Public Works Association when marking the horizontal route of any underground facility of the operator.

(8) Prior to or during excavation or demolition, if the marking of the horizontal route of any facility is removed or is no longer visible, the excavator shall stop excavation or demolition activities in the vicinity of the facility and shall notify the operator(s) to have the route remarked.

(9) If any contact with or damage to any pipe, cable, or its protective covering, or any other underground facility occurs, the excavator causing the contact or damage shall immediately notify the operator. Upon receiving notice, the operator shall send personnel to the location as soon as possible to effect temporary or permanent repair of the contact or damage. Until such time as the contact or damage has been repaired, the excavator shall cease excavation or demolition activities that may cause further damage to such underground facility.

Section 4. Liability of the operator and excavator.

(1)(a) Any person who violates subsection (1) or subsection (4) of section 3a and subsequently, whether by himself or through his employees, contractors, subcontractors, or agents, performs excavation or demolition work which damages an underground facility of an operator shall be strictly liable to such operator for:

1. Any cost incurred by the operator in repairing or replacing a damaged facility.

2. Any damage to property or any injury or death to a person resulting from damaging the underground facility of the operator.

3. Any loss of revenue or loss of use resulting from such damage to an underground facility.

Any such person shall also indemnify the operator against any claim for personal injury, death, property damage, or service interruptions resulting from damage to the underground facility.

(b) If any excavator fails to discharge a duty imposed by the provisions of this act, excluding subsections (1) and (4) of section 3, such person shall be liable for the total sum of the losses to all parties involved as those cost are normally computed.

(c) Obtaining information as to the location of an underground facility from the operator as required by this act does not excuse any excavator from performing an excavation or demolition in a careful and prudent manner, based on accepted engineering and construction practices, nor does it excuse such excavator from liability for any damage or injury resulting from any excavation or demolition.

(d) When an excavator knows or should know of the presence of an underground facility, he shall make reasonable efforts to contact the person who owns or operates that facility prior to commencing an excavation or demolition, regardless of whether that person in an operator.

(3) If, after receiving proper notice an operator fails to discharge a duty imposed by the provisions of this act and an underground facility of such operator is damaged by an excavator who has complied with the provisions of this act, as a proximate result of the operator's failure to discharge such duty, such excavator shall not be liable for such damage.

Section 5. Penalties.--

(1) Any excavator who knowingly and willfully violates any provision of this act may be assessed a civil penalty of not more than $1,000 for the first violation and not more than $5,000 for any subsequent violation during a 12-month period. Such penalties are in addition to any other civil penalties that may be imposed.

(2) Any person who knowingly and willfully removes or otherwise destroys the stakes or other physical markings used to mark the horizontal route of an underground facility commits a misdemeanor.

(3) An action may be brought by the Attorney General, the State attorney, or the local or state agency which issued the permit to excavate for the enforcement of the civil penalty pursuant to this section. If any penalty is collected as a result of a civil suit brought by a state or local agency for the collection of such civil penalties, the penalty imposed shall be paid to the general fund of the agency. If more than one agency is involved in enforcement, the penalties imposed shall be apportioned among them by the court in a manner that will fairly offset the relative costs incurred by the state or local agencies, or both, in collection such penalties.

Section 6. Single-family residential property exemption.--The notification requirements provided in subsection (1) of section 5 do not apply to any excavation or demolition performed by the owner of single-family residential property when such excavation or demolition is made entirely on such land, provided there is no encroachment on any operator's right-of-way, easement, or permitted use.

Section 7. Emergency excavations or demolitions attempted; exception.--The provisions of this act do not apply to making an excavation or demolition during an emergency, provided the operator was notified at the earliest opportunity and all reasonable precautions had been taken to protect any underground facility. For the purposes of this act, "emergency" means an condition constituting a clear and present danger to life or property caused by the escape of any material or substance transported by means of an underground facility; any interruption of vital public service or communication caused by any break or defect in an operator's underground facility; or, in the case of the State Highway System or streets or roads maintained by a political subdivision, if the use of such highways, roads, streets, or other public ways is, in the judgement of duly authorized officials of the Department f Highway Safety and Motor Vehicles and the Department of Transportation or such political subdivision, impaired by an unforeseen occurrence which necessitates repair beginning immediately after such occurrence. The time requirements for notification provided in this act do not apply to emergency excavations.

Section 8. Applicability to existing law.--Nothing in this act shall be construed to:

(1) Constitute the establishment or enlargement of any rights to the use of real property or create an interest therein for the placement, construction, repair, maintenance, relocation, or excavation or demolition of any underground facility.

(2) Waive any right of a party having an interest in real property to charge any fee for the use regarding such property; or

(3) Preempt a governmental operator from reasonable regulation of its right-of-way.

Section 9. If any provision of this act or the application thereof to any person or circumstance is held invalid, the invalidity shall not affect other provisions or applications of the act which can be given effect without the invalid provision or application, and to this end the provisions of this act are declared severable.

The Gas Company

Robert L. Beauregard
Manager
Federal Regulatory Affairs

MEMO FOR THE RECORD
Date Sent: 6/24/94
Hearing Date: 6/23/94
Material Forwarded to Printer: 6/24/94

Southern California Gas Company
Washington Center
1001
G Street, N.W.
Sixth Floor East
Washington, D.C.
20001-4545

tel 202 662-1706
fax 202 293-2887
fax 202 293-2888

June 24, 1994

Chairman Philip R. Sharp
Energy and Power Subcommittee
331 Ford House Office Building
Washington, D.C. 20515

Dear Chairman Sharp:

Please consider the enclosed comments from a combination of California parties for inclusion in the record of your June 23rd, 1994 Energy and Power Subcommittee hearing on H.R. 4394.

The one page General Comments are offered by the Los Angeles Department of Water and Power, Pacific Bell, Pacific Gas and Electric, San Diego Gas and Electric, Southern California Edison Company, Southern California Gas Company, and Underground Service Alert of Southern California.

The separate set of multi-page comments to H.R. 4394 are offered by Southern California Gas Company and Underground Service Alert of Southern California. These comments suggest a series of suggested improvements to H.R. 4394.

Thank you for your consideration.

Sincerely,

Robert L. Beauregard

Robert L. Beauregard

RLB/abp

Enclosures

SHARP, JUNE 16 DRAFT HR 4394

***GENERAL COMMENTS SUPPORTED BY*:**

LOS ANGELES DEPARTMENT OF WATER AND POWER, PACIFIC BELL, PACIFIC GAS AND ELECTRIC, SAN DIEGO GAS AND ELECTRIC, SOUTHERN CALIFORNIA EDISON COMPANY, SOUTHERN CALIFORNIA GAS COMPANY *and* UNDERGROUND SERVICE ALERT OF SOUTHERN CALIFORNIA

As dedicated "one-call" professionals, protecting underground facilities in the most populous State in the Union, we fully recognize the benefit of creating minimum national standards to provide for the establishment of Statewide excavator notification systems.

We also believe that a very practical Federal approach will be required to adequately provide for the broad diversity of underground facility operators, and varied circumstances of excavation activity, potentially subject to this legislation. This Act should also consider that Federal, and State, codes mandating "emergency plans" and "damage prevention programs" are currently in place for certain operators.

Because of the all encompassing nature of this Act, we favor a law prudently constructed that neither favors, nor burdens, one type of underground operator, or excavator, over another. We also believe that existing "one-call" systems meeting the minimum requirements can, and should be, the choice of "models" for those States lacking them, and that existing systems fulfilling the minimum requirements be fully recognized as the "State" program defined in the Act.

We advocate a sensible law which recognizes certain carefully defined excavators must pre-notify "one-call" systems, and that certain carefully defined owner/operators must respond to that call. The most effective Federal legislation will provide sound, clear, and logical minimum guidelines allowing States the ability to tailor more specific requirements to regional needs. We conclude that regional flexibility, and local determination to go beyond Federal requirements, are paramount to a sound Act. Consequently, Federal focus should be limited to:

- A requirement that States legislate (within an acceptable time frame) a requirement for mandatory Statewide notification systems, in order to provide excavators a single statewide telephone number, to pre-notify underground facility operators of their intent, timing, and location of pending excavation activity,

- That excavators, as defined, must pre-notify these systems within sufficient, reasonable, and codified time parameters, to allow for operator response,

- That owner/operators of underground plant, as defined, must be members of, support public education of, and financially sustain these systems, and that, within an established excavator pre-notification period, all members, or their agent, must respond to, field locate, and mark the surface location of their facilities ahead of the excavation, or alternately provide facility information in another manner sufficient for the excavator to determine the involvement, or non-involvement, of subject operative underground facilities,

- That States must provide for sufficient and effective penalty and enforcement codes to promote operator and excavator compliance.

SOCAL GAS COMPANY AND USA SOUTH POSITION RE: HR 4394

GENERAL COMMENTS - SHARP, JUNE 16 DRAFT

SoCal Gas, and USA/SC having been involved in "one-call" sysyems for 18 years welcomes the opportunity to comment on HR 4394. We recognize the potential benefit of creating minimum national standards to provide for the establishment of Statewide excavator notification systems.

We also believe that a very practical Federal approach will be required to adequately provide for the broad diversity of underground facility operators, and varied circumstances of excavation activity, potentially subject to this legislation. This Act should also consider that Federal, and State, codes mandating "emergency plans" and "damage prevention programs" are currently in place for certain operators.

Because of the all encompassing nature of this Act, we favor a law prudently constructed that neither favors, nor burdens, one type of underground operator, or excavator, over another. We also believe that existing "one-call" systems meeting the minimum requirements can, and should be, the choice of "models" for those States lacking them, and that existing systems fulfilling the minimum requirements be fully recognized as the "State" program defined in the Act.

We advocate a sensible law which recognizes certain carefully defined excavators must pre-notify "one-call" systems, and that certain carefully defined owner/operators must respond to that call.

The most effective Federal legislation will provide sound, clear, and logical minimum guidelines allowing States the ability to tailor more specific requirements to regional needs. We conclude that regional flexibility, and local determination to go beyond Federal requirements, are paramount to a sound Act. Consequently, Federal focus should be limited to:

- A requirement that States legislate (within an acceptable time frame) a requirement for mandatory Statewide notification systems, in order to provide excavators a single statewide telephone number, to pre-notify underground facility operators of their intent, timing, and location of pending excavation activity,

- That excavators, as defined, must pre-notify these systems within sufficient, reasonable, and codified time parameters, to allow for operator response,

- That owner/operators of underground plant, as defined, must be members of, support public education of, and financially sustain these systems, and that, within an established excavator pre-notification period, all members, or their agent, must respond to, field locate, and mark the surface location of their facilities ahead of the excavation, or alternately provide facility information in another manner sufficient for the excavator to determine the involvement, or non-involvement, of subject operative underground facilities,

- That States must provide for sufficient and effective penalty and enforcement codes to promote operator and excavator compliance.

SPECIFIC COMMENTS - HR 4394, JUNE 16 "SHARP" DRAFT

SOUTHERN CALIFORNIA GAS CO. -USA/SC

COMMENT #1

SECTION 1. SHORT TITLE

Currently States: **This Act may be cited as the "Comprehensive One-Call Notification Act of 1994".**

We Support: **This Act may be cited as the "Uniform One-Call Notification Act of 1994".**

Reason: "Uniform" more accurately describes the intent that all states adhere to the same minimum guidelines. "Comprehensive" has a wide interpretive potential which can confuse and muddle the spirit and/or purpose of the Act.

COMMENT #2

LONG TITLE (Bottom of page 14 of June 16 Sharp Draft)

Currently States: **"A Bill to provide for the establishment of mandatory, comprehensive, Statewide one-call notification systems to protect natural gas and hazardous liquid pipelines and all other underground facilities from being damaged by any excavations, and for other purposes."**

We Support: **"A Bill to provide for the establishment of, and minimum standards for, mandatory State legislated, excavator pre- notification centers (commonly known as "One-Call" systems), for the purpose of protecting underground facilities from damage caused by pending excavation activity."**

Reason: The word "comprehensive" is not defined in the bill, and it leaves it's intent open to any interpretation.

It is also redundant, and unnecessary, to single out "natural gas and hazardous liquid pipelines" as they should be included with, and subject to, the same provisions as any other sub-surface facility coming under the definitions outlined in this Bill. There are existing Federal guidelines specific to these type of facilities.

In reality, it must be recognized that "all" underground facilities cannot be protected from "any excavations". Therefore, it seems more appropriate to set a fundamentally utilitarian, and comprehensible, foundation for the Bill's intended purpose. The current title language lacks consideration that "all" underground facilities are not owned solely by the kinds of "facility operators" normally capable of adhering to the provisions of this bill (referring to some individual property owners, farmers, businesses, sewer districts, communal organizations, churches, schools <public and private>, camp grounds, etc.), and there are no such exclusions provided in this Bill's definitions".

HR 4394

SOUTHERN CALIFORNIA GAS CO. - USA/SC

COMMENT #3

SECTION 2, FINDINGS, ITEM (3)

Currently States: ....."**the vast majority**" of accidents are caused by excavation activity known as "....third party damage;"

We concur with reservations.

Reason: Any implied suggestion that a federally mandated one-call Act will end these incidents is misleading. "One call" does not prevent all incidents. In California (where exists a mandatory excavator prior notice requirement) damages still occur due to excavators ignoring the law, by negligence after prior notice, and even after the operator has provided the correct location information.

In the case of good codes, total damage prevention is only obtainable through the additional use of efficient internal operator policy, practical code enforcement measures, and education. California code, being fairly new, is still being refined.

New avenues of enforcement at the State level are being enacted at this time, and education is a prime objective of our "One-Call" efforts. Excavator calls to the "center" are increasing in number, and damages to our plant are decreasing. We believe all States are capable of enacting similar effective laws to protect facilities.

HR 4394

SOUTHERN CALIFORNIA GAS CO. - USA/SC

COMMENT #4

SECTION 2. FINDINGS, ITEM (6)

Currently States: **".....in the case of interstate facilities, consumers in one State may pay for damages incurred in another State;"**

We support: **Deletion of this phrase, unless the Bill goes on to preclude interstate companies from continuing to allot costs outside the area in which they have incurred them.**

Reason: Interstate operators chose their route of installation, and are fully cognizant of the "one-call" conditions in those areas. This condition should not be confused with an internal obligation of protecting buried facilities in the areas they chose to install them, nor used as an excuse for unfairly allocating costs.

While we support a Federal "one-call" mandate, we strongly believe this effort should be balanced, and fair, to all parties.

Intrastate operators are, by far, the vast majority of underground plant operators, both in number and miles of plant. Just one intrastate operator, SoCal Gas Company for instance, operates approximately 80,000 miles of underground piping. This is more plant than many of the interstate operators combined. SoCal Gas recognizes a responsibility to staff, and locate field offices, sufficient to provide responses beyond the "call before you dig" requirements. These resources are in direct proximity and relation to our underground plant. It would seem reasonable to expect all underground plant operators to provide a similar level of plant security.

Also, the "Findings" fail to mention that intrastate operators would now be forced, under many provisions this Bill, to share in the costs of a "National Telephone System", information center, or exchange which will not benefit the intrastate operator, nor a local consumer utilizing a local operators product or service.

HR 4394

SOUTHERN CALIFORNIA GAS CO. - USA/SC

COMMENT #5

SECTION 2. FINDINGS, ITEM (9)

Currently States: **"....these standards should apply, without exception, to all excavators and to all facility operators;.."**

We Support: **".... these standards should apply, with certain specified exclusions, to all excavators and to all facility operators:.."**

Reason: The existing language supports only one conclusion: it strives to accomplish a practical impossibility, i.e. the absolute protection of all substructures anywhere in the USA from all damages caused by any and all excavation activity, conducted by any and all excavators.

Without any exclusions and definitions (the specifics of which we believe are best left to the States) the current language would include all homeowners, all government and military installations, all private enterprises of every description, all schools and all universities as "operators", and as "excavators".

The sole exception is for "emergency" excavations (Section 6, Item <8>) which, even if completed, must also be reported at the "earliest practical time" (current Draft language).

The current "Finding" relates a noble objective, it is just not "real world".

HR 4394

SOUTHERN CALIFORNIA GAS CO. - USA/SC

COMMENT #6

SECTION 2. FINDINGS, ITEM (10)

Currently States **".... these standards should produce one-call notification systems which are simple to use, with a single toll-free telephone number established which excavators must call to obtain information on the location of any type of underground facility anywhere in the United States."**

We Support: **.... these standards should cause to be established one-call notification systems which are simple to use and which will provide an effective means for excavators to pre-notify facility operators in the vicinity of planned excavations.**

Reason: The current statement is not a realistic goal. A practical consideration must be measured against the multitude of existing facilities, and the various quality, or existence of, substructure records. It could never be an attainable, efficient, or a cost effective "one-call" mission to accumulate from the many thousands of operators the data necessary to update, fulfill, and maintain the stated objective. Also, underground plant information can be internally very sensitive, and therefore, under some circumstances, must be held internally secure.

The current clause also dictates that information will be supplied for which, in many cases, no data exists. A large number of out-of-business operators, smaller companies, older companies, municipalities, private mobile home parks, military establishments, universities and others simply do not have any mapping, much less complete, modern, or updated mapping. The resources and effort required to satisfy this requirement would be tremendous, if not impossible, and not necessary to the "one-call" mission.

We realize the issue of one National telephone number is valid for those who have interstate operations. We can support a National number but believe, in order not to disrupt effective Statewide systems, this number should be a one-call referral number that automatically switches the caller to the appropriate State system.

We further believe that all costs associated with establishing and maintaining a National number, which benefits only interstate operators, be borne entirely by such operators that wish to subscribe.

One alternative to a national number may be a annual printing of a "one-call" directory to include all Statewide 800 telephone numbers of "one-call" centers and those telephone numbers of those operators having plant in areas where temporarily no "one call" system exists. These directories could be funded by the interstate operators, and operators of systems in non-coverage areas, and made available to excavators through their union halls, professional organizations, trade shows, existing one-calls, and individual operators.

HR 4394

SOUTHERN CALIFORNIA GAS CO. - USA/SC

COMMENT #7

SECTION 3, DEFINITIONS, ITEM (2) EXCAVATION

Currently states: ...."**excavation means any operation in which earth, rock, ..., or other material is removed by means of any mechanized tools or equipment , or any ....but shall not include the tilling of soil to 18 inches...".**

We support: **"Excavation" means any operation in which earth, rock, or other material in the ground is moved, removed, or otherwise displaced by means of tools, equipment, or explosives in any of the following ways: grading, trenching, digging, ditching, drilling, augering, tunneling, scraping, cable or pipe plowing and driving, or any other way.**

Reason: The limitation to "mechanized" jeopardizes existing "one-call" systems that extend to excavations conducted by hand tools. Some states, including California, have determined that sufficient damages are caused by hand tool excavating, thereby warranting the coverage of these tools.

Some of these damages may be unavoidable due to the nature of the facility, and the conditions of the surrounding soil. However, where no location is required, even more damages will be incurred. It is very critical that all digging implements (within the context of all other definitions in the Act or State code) be considered as falling under the scope of a "one-call" system.

Also, the agricultural exemption should be deleted because many critical underground facilities are located under areas subject to constant and repeated plowing. Erosion, and other factors, can change the amount of earth cover over these facilities.

HR 4394

SOUTHERN CALIFORNIA GAS CO. - USA/SC

COMMENT #8

SECTION 3, DEFINITIONS, ITEM (3) EXCAVATOR

Currently States: **"...."excavator" means a person who conducts excavation;"**

We believe these types of specific descriptions are best left to the States and believe any Federal definition must be clear and concise. The current description includes every person from those digging sand castles, or planting flowers in their backyard, to those mining ore.

Consequently, if a Federal definition is unavoidable, that code should consider that an:

...."excavator" means any person conducting an excavation on public or private property for the purposes of commercial enterprise, monetary profit, governmental activity, or contracted property improvements or alterations requiring a building permit that involves any disruption of the sub-surface, or any other business or professional endeavor wherein an excavation is planned.

"Excavator" does not mean a person, owner, or entity, conducting an excavation on their privately owned property, and over their own facility(s), or for other domestic or non-commercial purposes, unless it is known by the person/owner/operator that "Right of Way" passage has been granted to another facility on that same property. Persons not meeting the "excavator" description may utilize, and are encouraged to use, the State "one-call" system in the same manner as those "excavators" required to do so.

Reason: We believe these definitions are best left to the States involved because States have the burden of enforcement, and State "one-call" systems will and do provide the education.

Depending on density, type, and opportunity, the regional conditions can affect who, or what, should apply to these types of definitions (i.e. what is an excavation, who is an excavator, and who is an operator.).

HR 4394

SOUTHERN CALIFORNIA GAS CO. - USA/SC

COMMENT #9

SECTION 3, DEFINITIONS, ITEM (4) FACILITY OPERATOR

Currently states: **"...."facility operator" means any person who owns or operates an underground facility;"**

We support: **....."Facility operator" means any person who owns or operates an underground facility, except for any person who is the owner of real property wherein are located underground facilities for the purpose of furnishing services or materials only to himself or occupants of such property.**

Reason: "One-call" system codes should exclude homeowners, businesses, or entities that are conducting their own non-commercial, or personal, activities that pose little to no threat to the public.

COMMENT #10

SECTION 5, STATE PROGRAMS, (a)

Currently states: ....."**Each State shall....... provide for the establishment of a comprehensive statewide one-call notification program.........to protect all underground facilities from damage due to any excavation."**

We support: **Each State shall, within 3 years after the effective date of this Act, provide for legislation to establish statewide one-call excavator pre-notification systems in accordance with this Act.**

Reason: The current language is all encompassing, and ambiguous. Any workable Act will require some practical exclusions for subject activities and subject facilities in order not to include unwanted, useless, or unenforceable provisions.

Also, the current language infers that only State governments will establish "one-call" programs. We support an Act that provides for continuation of successful facility owner/ operated "non-profit" one-call systems.

HR 4394

SOUTHERN CALIFORNIA GAS CO. - USA/SC

COMMENT #11

SECTION 5, STATE PROGRAMS, (b)

Currently grants: ...."$12,000,000" in public funds **"to States for use in establishing one-call programs....".**

We support: **Deletion of this entire subsection. We support the concept of State legislation to mandate State systems. We do not support a State "operated" one-call system. It is our position that the owner/operators should completely fund and be responsible for the protection of their underground plant, and that this burden should not be shifted onto the State government, or onto State or Federal tax payers.**

Reasons: States should not seek public funds for what are basically private corporate issues, i.e. the protection of underground plant from outside digging activity. These costs have been, and should continue to be, borne by the entities that installed the underground facilities.

Federal grants to aid States in developing systems to protect private facilities is a misdirection of public funds.

Operators should subsidize their own damage prevention measures. This is the accepted, and practical, arrangement of existing "one-call" systems.

84-088 O - 94 - 8

HR 4394

SOUTHERN CALIFORNIA GAS CO. - USA/SC

COMMENT #12

SECTION 6, ELEMENTS OF STATE PROGRAMS, ITEM (1) (B), AND (C) (i) through (iv)

(1) (B) Currently states: **(Programs shall)...."apply to all excavators and to all facility operators, ...**

We support: **(Programs shall)....apply to all excavators as defined by the State, and to all facility operators as defined by minimum Federal and/or State guidelines and ,**

Reason: We believe that excavator definitions are regionally sensitive, and must be designed around local conditions. We also recognize that any interpretation of underground "facility operators" must, without further definition, undoubtedly and incontestably, include operators of critical services, or hazardous liquids or gases.

However, not all owners of all facilities need to belong or adhere to a "one-call" system. Either because their facilities are entirely on property completely controlled and unique to them, or because the requirement to have them belong is more burdensome than otherwise (i.e. non-pressurized and non-locatable sewer or drain system operators).

(C), (i) through (iv) Currently dictates finite detail of internal "one-call" administrative requirements.

We support: **Deletion of these sub-sections.**

Reason: We contend that this detail is unnecessary micro management of internal State established systems. The key should be in obtaining the goal. We believe specific office administration functions should be determined by the "one-call" members and staffs.

HR 4394

SOUTHERN CALIFORNIA GAS CO. - USA/SC

COMMENT #13

SECTION #6, ELEMENTS OF STATE PROGRAM, ITEM 1 (G)

Currently states: ....."**provide for 24-hour coverage for emergency response;**"

We support: **Complete elimination of this clause.**

Reason: This clause is not supportive of pre- notification of planned excavations, the underlying goal of this legislation. Off-hours coverage should be left to the discretion of the State "one-call" system.

This Bill's pre-notification requirement provides excavators plenty of opportunity to pre-notify systems of planned activity during normal working hours, and normal business days. Further, California's "one-call" procedures provide for "Short Notice" calls to accommodate unusual circumstances. While "Short Notice" is less than 2 days, it is still during normal business hours.

If the intent is to take "emergency" type digs, even this bill wisely (as is the case with existing California code) omits the requirement of pre-notification for emergencies. True "emergency" conditions require immediate action, and in the vast majority of these cases no pre-determination of the need to dig, or the location of such an excavation, can be made before the responding operator is on the "emergency" site.

Operators of facilities, capable of being the catalyst for such "emergencies", are the only responsible excavators responding to such incidents. They must make the on-site decision to dig and rectify the situation immediately. They do not have the latitude to do otherwise. Waiting for other "one-call" members to respond, to an after normal hours field locate, could spell the difference between avoiding a disaster, and causing one.

A 24 hour service, is extremely expensive. To require all utility operators, and ultimately their customers, to bear this expense goes beyond damage avoidance into the realm of providing a free, and very open to abuse, service for excavators convenience. A 24 hour service circumvents, and undermines, the pre-notification requirement.

Costs and or liability associated with providing these excavator (and non-mutual benefit) services would be unfairly borne by the all member operators, and ultimately, their customers.

This clause also does not take into account that some utility operators already staff for 24 hour operations which are mandated and necessary for the protection of the public, and it's particularly unique plant. They already accept after hours "emergency" type facility locate requests together with all other potentially critical response calls. These requests are evaluated by highly qualified and trained personnel in order to determine the level of risk or involvement. Immediate remedial action can only be properly determined by these trained personnel speaking directly to the caller.

To assume that an "emergency" function could be taken on by a "one-call" does not take into account the density, or variety, of underground plant in some locals, the levels of specific expertise that can be involved to properly screen such calls, or the criticality of a timely response based on correctly interpreted circumstances.

Added risk is probable if the general public is lead to believe that a "one-call" system is some sort of "emergency" response entitiy, or central service call center. Incidents of potentially very hazardous delays, and misunderstandings are likely.

Further, many companies do not have the staffing, need, available funds, operational requirement, nor safety sensitive plant to provide 24 hour services. If a one-call service (to which all operators must belong) takes a locate request for after normal hours, then it stands to reason that all operators will be forced to respond during after normal business hours. All operators would eventually be forced to respond because a "one-call" center cannot for very long provide dual levels of service (i.e. some members respond, others cannot) to a single excavating community.

HR 4394

SOUTHERN CALIFORNIA GAS CO. - USA/SC

COMMENT #14

SECTION 6, ELEMENTS OF STATE PROGRAMS, ITEM (5), LINE 11 AND 12

Currently states: **"require facility operators to provide for locating...........within no more than 2 business days after notification ...."**

We support: **" require facility operators to provide for locating........within 2 business days after notification, or no later than 14 calender days after notification, or ahead of the stated time of the excavation, whichever is later,**

Reason: The 2 day requirement needs to be coupled with the outside limit of notification to allow operators the latitude to mark their facilities any time after notification providing such marking will be ahead of the stated digging activity timing. The outside notification window should consider that variously observed facility operator holidays can affect what constitutes 10 business days. Therefore, a 14 day period accomplishes the intended goal (i.e. 2 weeks).

Also, Currently states: **"....., and to monitor such excavation as appropriate;"**

We Support: **Deletion of the entire phrase.**

Reason: The wording is ambiguous. This wording shifts the burden of correct, and undeniable excavator responsibility which rightly dictates that the excavator (as the intruding party) is to perform in a responsible manner while conducting activities near an otherwise safe, secure, and in place underground plant location.

Considering the numbers of requests (350,000 annually for SoCal Gas alone), a requirement to monitor any specific excavation activity, after marking the location should be at the complete discretion of the operator.

The word "appropriate", being pointedly undefined, will be subject to interpretation relating to circumstances beyond the control, or knowledge, of the operator.

HR 4394

SOUTHERN CALIFORNIA GAS CO. - USA/SC

COMMENT #15

SECTION 6, ELEMENTS OF STATE PROGRAMS, ITEM (7)

Currently states: **"provide for procedures to improve mapping of underground facilities, including adding to appropriate mapping systems any information about previously unknown underground facility locations when such information is discovered;"**

We support: **Delete this entire provision.**

Reason: This could only be viable, or possible, if the State was responsible for maintaining all operator mapping. "One-call" systems would also be inundated with unnecessary activity, and huge expenses, trying to fulfill this clause.

Some "one-call" systems are working toward more uniform base, or reference, mapping in order that members may better report the underground plant locations, but only where the need exists, and only funded by the involved operators. To mandate this activity across the board is not conceivable, or practical.

To further suggest that a "one-call" system should update mapping systems, regarding "previously unknown facilities", is asking a "one-call" system to assume an operators responsibility to know where it's plant is located. Liability for questions regarding accuracy, to whom any newly located facility belongs, who will locate it, and additional non-damage prevention functions arise with this issue.

HR 4394

SOUTHERN CALIFORNIA GAS CO. - USA/SC

COMMENT #16

SECTION 7, PENALTIES AND ENFORCEMENT (a) and (b).

(a) Currently provides: **Up to "$500,000 "for "administrative or civil penalties".**

We Support: **Lowering the amount of liability to no more than $10,000 for negligent offenses, and $50,000 per incident for knowing and willful violations.**

Reason: We believe the focus of enforcement should be on furthering education, and promoting safe practices. Punitive burdens large enough to put normally responsible small excavators, or operators, out of business are not conducive to building the kinds of cooperative relationships that should otherwise be developed between enforcement agencies, owners, and excavators.

Given a capable enforcement vehicle, a $10,000 or $50,000 fine is more than sufficient to gain attention, and adherence, to the Act.

(b) Currently provides: **"If a violation results in damage.... resulting in death, serious bodily harm, ...or damage...... exceeding $50,000......penalties may be increased, ...imprisonment may be assessed."**

We support: **Complete deletion of (b).**

Reason: These punitive measures are extreme and redundant. Existing liability and criminal codes should suffice to recover damages, or prosecute individuals.

COMMENT #17

SECTION 9, ALTERNATE FORM OF STATE PROGRAM

This section needs refining to answer the following questions:

1) Does an alternate plan remain, or become, effective in the 6 month period of Secretary review?

2) Can existing systems qualify under the same provisions?

3) If denied, is there an appeal process?

Gas Processors Association

PRESIDENT JOHN EHLERS
VALERO HYDROCARBONS, L.P., SAN ANTONIO

PRESIDENT-ELECT DON DUNLAP
CHEVRON/WARREN, TULSA

VICE-PRESIDENT W. H. SPEAKER, JR.
MOBIL R & D CORP., DALLAS

VICE-PRESIDENT JOE BECRAFT
TRANSOK, INC., TULSA

VICE-PRESIDENT DON P. AMSDEN
UNION PACIFIC RESOURCES CO., FORT WORTH

6526 EAST 60TH ST.
TULSA, OKLAHOMA 74145
TELEPHONE (918) 493-3872
FAX (918) 493-3875

MARK F. SUTTON, EXECUTIVE DIRECTOR
JANET D. HERBERT, DIRECTOR, RESEARCH AND GOVERNMENTAL AFFAIRS
RONALD G. BRUNNER, DIRECTOR, TECHNICAL SERVICES

P.O. BOX 35584
TULSA, OKLAHOMA 74153

Testimony of the Gas Processors Association submitted for the record of the hearing on June 23, 1994 of the Energy and Power Subcommittee of the House Energy and Commerce Committee

Mr. Chairman, I am Mark Sutton, Executive Director of the Gas Processors Association ("GPA"). First, I would like to thank the Chairman on behalf of GPA for holding the hearing on June 23, 1994 on the very important issue of pipeline safety. GPA, headquartered in Tulsa, Oklahoma, is a trade association comprised of approximately 150 member companies, each of which is engaged in at least one of several activities involving production, gathering, treating, processing, transportation, and storage of natural gas and natural gas liquids ("NGLs"). Collectively, GPA member companies operate approximately 700 natural gas processing plants in the United States; process about 90% of all natural gas; and account for approximately 90% of all natural gas liquids produced in the United States. Although GPA membership includes most of the major oil and gas companies and a number of international companies, about 75% of the U.S. GPA members are small-to-medium sized independent, non-integrated producers and processors of natural gas.

I would like to make some brief comments about the two pieces of legislation under consideration, H.R. 4394, "The Comprehensive One Call Notification Act of 1994," and H.R. 4616, "The Natural Gas Pipeline Safety Improvement Act of 1994."

Comprehensive One-Call Notification Act

The Gas Processors Association ("GPA") supports comprehensive One-Call systems. Most of the members belong voluntarily to One-Call systems wherever they exist. Mandating 100% participation of all underground facility operators (public and private) and mandatory use by excavators would maximize the effectiveness of the program and reduce accidental damage to underground facilities during excavations. Therefore, we feel legislation concerning this subject would be in the best interest of all concerned.

While we believe in the concept of the national One-Call system, it is our opinion that the Federal Government should not be involved with state One-Call systems other than to mandate their existence, to require participation in all states, and to utilize the national One-Call number to promote public awareness and information concerning the state systems.

Most states already have privately run One-Call systems (see attached One-Call Directory list) and, in our opinion, they are very effective. However, some states have multiple One-Call systems. If a facility operator belongs to one, and an excavator calls another, the operator is not notified. In these cases, legislation needs to be mandated to allow only one One-Call system per state or require systems within the same state to notify each other when they receive a notification of intent to excavate.

Beyond our general opinions on One-Call systems, I would like to make some specific comments on particular parts of the bill. First, a time frame of three (3) years is recommended for maintaining notices of intent to excavate where no incidents or accidents are involved and seven (7) years for those notices of intent involving incidents or accidents.

Second, the requirement to provide One-Call system information at the time of issuance of excavation or building permits needs clarification. Does this mean that when an excavator applies for a permit he will be handed One-Call information, or is it suggesting that, in the future, all excavators will require a permit and must be informed of the One-Call system and its regulations at the time of permitting? If the latter is true, we will not support issuing excavation permits because this process could become very cumbersome and costly, causing extensive scheduling delays. Requiring a one or two year excavator equipment operator license instead of issuing permits would be a more appropriate and less costly action, placing the financial burden on the excavator.

Third, the imposition of fees to operate the One-Call system should be shared by both excavator and facility operator. In the present privately owned system, the operator already has the burden of membership fees to support the system and supply manpower to the excavation site with no reimbursement.

With the exception of these few issues, GPA supports the enactment of a new Comprehensive One-Call system.

Natural Gas Pipeline Safety Act

On the whole, the transportation of gas through pipelines is one of the safest forms of energy transportation, if not the safest. While GPA operators fully agree with the need for the National One-Call Notifications system, we feel that the reforms to the pipeline safety system encompassed in the "Natural Gas Pipeline Safety Act" are not necessary. Most of its requirements are duplications or poor replacements of already existing Department of Transportation rules. Also, we feel it should not be the place of Congress to mandate exact solutions. The regulatory agencies should maintain flexibility to provide applicable solutions to problems in various situations.

Beyond the general comments, I would like to make several points about specific parts of the proposed legislation. First, the Federal Energy Regulatory Commission ("FERC") should not become involved with pipeline safety issues, which are already under the jurisdiction of the Office of Pipeline Safety ("OPS") in the Department of Transportation. Placing jurisdiction in two executive departments will only create confusing, possibly conflicting, regulation. Further, if the views of the two agencies conflict, regulatory gridlock may result which would work against the intentions of the legislation and the safety of the public.

Second, increasing the width of pipeline Right of Way ("ROW") has little bearing on the subsequent damage to the pipeline and on the land related damage. Careless excavators may strike a pipeline regardless of ROW width. The real issue here is how wide is wide enough. This is a design issue that must be based on relevant factors.

Third, although increasing public awareness is invaluable in the effort of pipeline safety, the provisions in the bill already exist in OPS regulations. There are current mandated regulations in the Code of Federal Regulation 49 §§ 192.614 and 195.440 covering public awareness programs of gas and liquid pipelines. It remains to be seen how the changes proposed would enhance the current public awareness system. Further, additional regulations will likely create unnecessary burdens for operators without enhancing safety.

Finally, remotely controlled valves may be helpful in isolating a pipeline, but only <u>after</u> rupture. A more effective solution is the prevention of pipeline failures through public awareness and participation in one-call systems.

Once again, I would like to thank the chairman for holding the hearing and allowing GPA to submit testimony for the record. I would be happy to work with the committee in the future to clarify our position and to assist the committee in any way possible with these important pieces of legislation.

ONE-CALL DIRECTORY

ALABAMA
Alabama Line Location Center (800) 292-8525

ALASKA
Locate Call Center of Alaska Inc.
(907) 278-3121

ARIZONA
Arizona Blue Stake Inc. (602) 263-1100
Tucson Blue Stake Center (602) 792-2211

ARKANSAS
Arkansas One Call System Inc. (800) 482-8998

CALIFORNIA
Underground Service Alert North
(800) 642-2444
Underground Service Alert South
(800) 422-4133

COLORADO
Utility Notification Center of Colorado
(800) 922-1987

CONNECTICUT
Call Before You Dig (800) 922-4455

DISTRICT OF COLUMBIA
Miss Utility (800) 257-7777

DELAWARE
Miss Utility of Delmarva (800) 282-8555

FLORIDA
Florida One Call (800) 432-4770
No Cuts (800) 542-0088

GEORGIA
Utilities Protection Center Inc. (800) 282-7411

IDAHO
Palouse Empire Underground Coordinating Council (208) 882-1794
Utilities Underground Location Center
(800) 424-5555
Panhandle Utility Coordinating Council
(208) 765-4272
Dig Line (800) 342-1585
One Call Concepts (800) 626-4950

ILLINOIS
Julie Inc. (800) 892-0123
Digger (Chicago Utility Alert Network)
(312) 744-7000

INDIANA
Indiana Underground Plant Protection Services
(800) 382-5544

IOWA
Underground Plant Location Service Inc.
(800) 292-8989

KANSAS
Kansas One-Call Center (800) DIG-SAFE

KENTUCKY
Kentucky Underground Protection Inc.
(800) 752-6007

LOUISIANA
Dottie System Inc. (800) 272-3020

MAINE
Dig Safe - Maine (800) 225-4977

MARYLAND
Miss Utility (800) 257-7777
Miss Utility of Delmarva (800) 282-8555

MASSACHUSETTS
Dig Safe - Massachusetts (800) 322-4844

MICHIGAN
Miss Dig Utility Communication System
(800) 482-7171

MINNESOTA
Gopher State One Call (800) 252-1166

MISSISSIPPI
Mississippi One-Call System Inc.
(800) 227-6477

MISSOURI
Missouri One Call System Inc. (800) 244-7483

MONTANA
Utilities Underground Location Center
(800) 424-5555

NEBRASKA
Nebraska Underground Hotline Inc.
(800) 642-8434
Lincoln One-Call (402) 477-0547

NEVADA
USA North and the State of Nevada
(800) 227-2600

NEW HAMPSHIRE
Dig Safe - New Hampshire (800) 225-4977

NEW JERSEY
Garden State Underground Plant Location Service (800) 272-1000

NEW MEXICO
New Mexico One Call System Inc.
(505) 260-1990

NEW YORK
Underground Facilities Protection Organization
(800) 962-7962
Utility Call Center (516) 661-6000
New York City One Call Center
(800) 272-4480

NORTH CAROLINA
Utilities Locating Co. Inc. (800) 632-4949

OHIO
Ohio Utilities Protection Service
(800) 362-2764

OKLAHOMA
Call Okie (800) 522-6543

OREGON
Utilities Underground Location
(800) 424-5555
Douglas Utilities Coordinating Council
(503) 672-5719
Josephine Utilities Coordinating Council
(503) 476-6676
Rougue Basin Utility Coordinating Council
(503) 779-6676
Central Oregon Underground Utilities Coordinating Council (800) 488-0675
Malheur Utility Coordinating Council
(503) 889-2468
Utilities Notification Center
(503) 246-6699
Curry Utilities Coordinating Council
(503) 469-2114

PENNSYLVANIA
Pennsylvania One Call System Inc.
(800) 242-1776

RHODE ISLAND
Dig Safe - Rhode Island (800) 225-4977

SOUTH CAROLINA
Palmetto Utility Protection Service Inc.
(800) 922-0983

TENNESSEE
Tennessee One-Call System
(800) 351-1111

TEXAS
Texas One Call System (800) 245-4545
Austin Area Utility Coordinating Council
(512) 499-7161
Texas Excavation Safety System
(800) 344-8377
Lone Star Notification Center
(713) 223-4567

UTAH
Blue Stakes Location Center
(800) 662-4111

VERMONT
Dig Safe - Vermont (800) 225-4977

VIRGINIA
Miss Utility of Virginia (800) 552-7001
Miss Utility (800) 257-7777
Miss Utility of Delmarva (800) 441-8355

WASHINGTON
Utilities Underground Location Center
(800) 424-5555
Grays Harbor & Pacific County Utility Coordinating Council (206) 532-3550
Utilities Council of Cowlitz County
(206) 425-2506
Chelan-Douglas Utilities Coordinating Council
(509) 663-6111
Upper Yakima County Underground Utilities Council (509) 248-0202
Inland Empire Utility Coordinating Council
(509) 456-8000
Palouse Empire Utilities Coordinating Council
(208) 882-1974
Utilities Notification Center
(206) 696-4848

WEST VIRGINIA
Miss Utility of West Virginia Inc.
(800) 245-4848

WISCONSIN
Diggers Hotline Inc. (800) 242-8511

WYOMING
West Park Utility Coordinating Council
(307) 587-4800
Call-In Dig-In Safety Council
(307) 682-9811
Fremont County Utility Coordinating Council
(307) 856-7555
Central Wyoming Utilities Coordinating Council (307) 265-5252
Sweetwater County Underground Coordinating Council (307) 362-8888
Carbon County Utility Coordinating Council
(307) 324-6666
Albany County Utility Coordinating Council
(307) 742-3615
Southeast Wyoming Utilties Coordinating Council (307) 638-6666
Coordinating Council (800) 562-5561
Wyoming One-Call (800) 348-1030

National Cable Television Association

Decker Anstrom
President &
Chief Executive Officer

1724 Massachusetts Avenue, Northwest
Washington, D.C. 20036-1969
202 775-3651 Fax 202 775-3695

June 22, 1994

The Honorable Philip R. Sharp
Chairman
Subcommittee on Energy and Power
Committee on Energy and Commerce
U.S. House of Representatives
Washington, D.C. 20515

Dear Mr. Chairman:

Thank you for giving the National Cable Television Association the opportunity to comment on H.R. 4394, the Comprehensive One-Call Notification Act of 1994. NCTA appreciates your efforts to protect both the public and owners of underground facilities – including cable television companies – from unnecessary disruptions in service.

One-call programs are desirable if (1) participation by all excavators, including contractors and local municipalities, is mandatory; (2) cuts in marked lines result in meaningful penalties, so that it is not cheaper for contractors to dig rather than wait at the site for underground facilities to be flagged; (3) the program is administered in an economical fashion; and (4) there are provisions to allow rapid restoration of service that has been interrupted by others. As such, the cable industry wishes to highlight its concern about several provisions in the bill which may need further refinement:

Section 5. Some of the best existing one-call programs are run by one of the large utilities in a state or by a co-op set up by participating companies. If a privately run system meets the federal government's minimum standards, it should not be necessary to create a new public bureaucracy to supplant a program which already works. Participation could be made mandatory, enforcement power could be vested in the state, and the program itself could continue to be administered privately.

The Honorable Philip R. Sharp
June 22, 1994
Page Two

Section 6. NCTA's general concern here is that no undue burdens be placed on cable operators beyond those which are necessary to the proper functioning of a one-call program. For example, subsection (3) provides for notification at least two business days before excavation begins in order to give other facility operators time to mark their facilities. This is a reasonable requirement under normal circumstances, but it does not deal with exigencies that may arise. Subsection (8) does provide for emergency timeliness waivers, but only in the case of "a threat to the environment or to public health and safety." It is not clear that this would apply to a loss or serious degradation of cable service.

Waiting at least "two business days" for restoration of cable service is not an acceptable prospect. Indeed, the FCC's customer service rules require attention to such matters within 24 hours unless there has been a more general disaster. We would suggest that either the definition of "emergency" be expanded to include a service outage by a facilities operator, or that a distinction be made between construction and repair. In the latter case, the excavator would provide notification as soon as possible but would be authorized to do whatever digging is needed to make repairs – at its own risk.

The requirement in subsection (9) that there be a schedule of fees to support a one-call program is understandable. However, cable operators are subject to strict rate regulation at the Federal Communications Commission. The legislative history of H.R. 4394 needs to make clear that such fees are a mandatory levy which can be passed through to subscribers like other costs which are beyond the control of a cable operator.

Section 7. Two aspects of this penalty section are of concern. First, the monetary penalties in subsection (a) need to be focused on excavators who are unwilling to wait for the one-call system to work before digging. As facilities operators, cable companies rarely excavate to the depth where gas, sewer, and water pipes are laid. Moreover, much of a cable system's facilities consists of individual drops run through easements to private homes, not trunk lines using public rights of way. Thus, the chance of cable excavators causing serious damage to persons or property is low. Concomitantly, cable companies are usually the victims of cuts by other excavators. If the dollar amounts of penalties in the bill remain as high as they are now, legislative history is needed to guide the assessment of penalties for infractions which cause little more than personal inconvenience.

The Honorable Philip R. Sharp
June 22, 1994
Page Three

Second, subsection (e) permits a state to revoke the license of a habitual violator. We realize that this remedy is intended to be used only in egregious cases, and it should be directed at contractors who, once on site, find it cheaper to dig rather than wait for facilities operators to mark their lines.

The broad franchise revocation language also causes cable companies concern since our business is different from other industries. For example, in most states, cable television franchises are awarded at the local level by cities and counties. Moreover, renewal and revocation of cable franchises are governed by federal law. If a cable operator is guilty of unwarranted service disruptions to other facilities, this can be a factor in any local renewal or revocation proceeding since compliance with one-call programs is usually covered by local franchises. NCTA does not think the existing regulatory scheme for awarding cable franchises should be superseded or that penalties aimed at lifting the permits of errant excavators should be extended to revoking the licenses of facilities operators.

Thank you for the chance to comment on H.R. 4394 for the record. I will be pleased to send a representative to your subcommittee's informal work session on Friday, June 24, regarding further changes to the bill.

Sincerely,

Decker Anstrom
President & CEO

Statement of A.E. "Ted" Bragger
President
National Utility Contractors Association

My name is Ted Bragger, and I am President of the A.E. Bragger Construction Co. located in Warwick, Rhode Island. Our family business was started in 1927, and I began working for the company in 1935 as a water boy. For six decades now, I have performed excavation work necessary for the installation, replacement, and repair of underground water treatment and delivery systems and wastewater collection and treatment systems. I still spend a couple of days each month on the seat of a backhoe. I also serve as the elected President of the National Utility Contractors Association (NUCA), which represents the women and men who construct water, sewer, gas, and other underground utilities and supply the equipment, materials, and services for construction. Our industry is a major stakeholder in the one-call process because we work where the action occurs, every single day.

I can assure you that great strides have been made over the past decades in locating and marking underground infrastructure. In the old days, we took it upon ourselves to contact utilities individually to determine if their lines were located near our jobsites. Meanwhile, the utilities, at great expense, would have to send a so-called "line walker" to the jobsite to monitor our excavation work until completion. Modern one-call systems have prevented scores of accidents and saved everybody a lot of time and money. Many of our chapters and members from around the country have worked at the local level to craft effective one-call notification systems that are accessible and easy to use. We think most of them do a good job and claim many one-call systems as active NUCA members.

NUCA formed a One-Call Task Force of seven contractors to analyze H.R. 4394 following its introduction last month. The Task Force has examined the June 16, 1994, draft substitute to H.R. 4394 and asks for your consideration of the following comments and recommendations, which we will update as additional feedback is received.

Section 1. Short Title

The title is appropriate because it is simple and recognizes the broad array of one-call notification topics contained in the bill. Also noteworthy is the fact that the measure would establish uniform national standards.

Section 2. Findings

We generally concur with the substance of the congressional findings and do not recommend any changes to findings (2), (5), (6), (8), (9), and (10).

Finding (1). Underground facilities have existed in the United States since the birth of the nation. Nevertheless, the construction of underground facilities skyrocketed beginning in the 1950s due to technological progress, accelerated economic growth, and the relocation of aerial facilities.

> Proposed Change to Finding (1):
>
> "(1) since the 1950s, ~~steadily increasing development of infrastructure has resulted in the construction of underground facilities throughout~~ the United States *has experienced unprecedented growth in the construction of underground facilitie*s, including . . . ."

Finding (3). Third-party damage results from excavator failure to contact one-call notification programs, improperly marked facilities, unmarked facilities, the absence of one-call notification programs, nonparticipation of facility owners in established one-call notification programs, equipment operator error, and other unforeseen events. It is important to briefly describe the manageable causes of third-party damage in the findings section so that they can be addressed in Section 6 (Elements of State Program), Section 7 (Penalties and Enforcement), and Section 8 (Department of Transportation--Model Program).

Proposed Change to Finding (3):

"(3) *most* ~~the vast majority of~~ accidents involving these facilities are caused by nearby excavation, demolition, or tunneling activities, known as third-party damage, *which commonly results from excavator failure to contact one-call notification systems, improperly marked facilities, and the nonparticipation of some facility owners in one-call notification programs*;

Finding (4). Accidents resulting from third-party damage present a clear and serious risk to public safety, employee safety, the natural environment, and property. The number of accidents can be substantially reduced but never eliminated due to human error and the fact that excavation is inherently dangerous work.

Proposed change to Finding (4):

"(4) accidents resulting from third-party damage present ~~an unnecessary~~ *a serious but manageable* risk to public safety, *property,* and the environment;

Finding (7). In many instances, the private construction community has led the fight for the creation and/or strengthening of one-call notification systems.

Proposed change to Finding (7):

"(7) to prevent third-party damage, the owners of some underground facilities *and concerned excavators* have initiated . . . ."

Section 3. Definitions

At this time we have comments but no proposed changes pertaining to Section 3.

First, we note that the definition of "excavation" has been modified to include only mechanized tools or equipment. It is our understanding that a substantial number of facility cuts are caused by hand tools.

Second, the definition of "excavation" exempts agricultural tilling to a depth of 18 inches or less. In a similar vein, a minority of our members have suggested an exemption from the definition for grading or scraping to a depth of 12 inches or less below existing grade. In theory, underground utilities are not located in this zone. Our official position at this time is that even shallow excavations present a safety threat to our equipment operators and other workers.

Third, we strongly support the new, simplified definition of "facility operator." The definition contained in H.R. 4394 as introduced excluded manufacturers, retailers, hospitals, and other entities that own substantial underground facilities.

Section 4. Nationwide Toll-Free Number System

The creation of a nationwide toll-free number system is laudable as long as it is feasible and easy to use. We support Section 4 as currently drafted and are particularly pleased that the Secretary of Transportation will consult with all the stakeholders before establishing this system.

Section 2(10) calls for a single toll-free number that can be used anywhere in the United States. If a single number is implemented, it must be capable of routing incoming calls from excavators to any state where excavation is planned (e.g., if a New Jersey contractor calls to provide notice of a planned excavation in Ohio, the call must be routed to Ohio).

Section 5. State Programs

The three-year time frame for state compliance in Section 5(a) is reasonable. In addition, we strongly support the authorization of grants to states in Section 5(b). Seed money for small facility owners to facilitate universal participation in one-call notification systems is particularly important.

We oppose the withholding of federal highway funds as a penalty for state noncompliance with Section 5(a), as set forth in Section 5(c). Instead, we recommend that Section 5(c) be deleted. If after three years following enactment of this measure, a state has not established the requisite one-call notification program, this issue can be revisited. If the Subcommittee insists that a federal sanction against unresponsive states be included in this measure, we encourage the Subcommittee to propose a sanction that does not also threaten the contractors who build safer roads and support the heart of the bill--its call for universal facility owner participation.

Section 6. Elements of State Program

We have a number of proposed changes pertaining to this section. We support subsections (1)(A), (1)(B), (1)(C), (1)(D), (1)(E), (1)(F), (2), (6), and (8), as currently drafted.

Subsection (1)(G). We support this provision to provide for 24-hour coverage for emergency response every day.

Proposed Change to Subsection (1)(G):

"(G) provide for *7-day,* 24-hour coverage for emergency response;"

Subsection (3). We support the requirement that excavators contact one-call notification systems at least 2 business days before excavation begins. This notice requirement is consistent with notice requirements in many one-call systems. We have a difference of opinion over whether 10 business days or 20 business days should be allowed before excavation must begin. Those who favor only 10 days argue that markings disappear when excavation is delayed. Those who favor a 20-day window point out that 10 days is not enough to accommodate unanticipated weather conditions, delayed delivery of supplies and materials, and other factors that prevent the timely commencement of excavation. It may be best to decide this issue on a state-by-state basis. Therefore, we recommend the more expansive timeframe for the minimum national standard.

> Proposed Change to Subsection (3):
>
> "(3) require that any person conducting an excavation must contact the one-call notification system at least 2 business days, and not more than ~~10~~ *20* business days, before excavation begins;

Subsection (4). We do not oppose the substance of this subsection unless its intent is to require that excavators be informed of excavation procedures every time they contact a one-call notification system. This would be impractical, except in the case of first-time callers. Moreover, this subsection may be redundant because Subsection (2) already stipulates that excavators be informed of state program requirements related to excavations.

Subsection (5). Our members favor inclusion of a national standard governing the accuracy of markings. For example, the location of an underground facility must be marked and identified within 18 inches on a horizontal plane from both sides and a vertical plane from the top of the facility. We are still developing a position and will forward proposed language shortly.

Subsection (7). This subsection requires states to improve the mapping of underground facilities, which is an important component in the drive to reduce third-party accidents. We recommend that the Subcommittee go one step further and mandate that facility owners, during the installation of underground plant, take necessary measures to ensure that their structures can be reasonably located at a later date. An example would be the placement of tracer wire above concrete pipe or non-metallic casing for fiber optic and gas lines.

Subsection (9). We object to this subsection as currently drafted. The costs of establishing, maintaining, and operating the State programs should be borne by the facility operators who own and are responsible for the underground facilities, as required in H.R. 4394 as introduced. One-call notification systems must be designed to encourage all excavators to use the system.

Proposed Change to Subsection (9):

"(9) provide for a fair and appropriate schedule of fees *to be imposed on facility operators* to cover the costs of establishing, maintaining, and operating the State program; and. . ."

Subsection (10). Our One-Call Task Force has not taken a position on this subsection.

Section 7. Penalties and Enforcement

The proposed penalties in Subsections 7(a) and 7(b) are too severe. The penalties should apply to willful violations only, especially in the case of Subsection 7(b), which prescribes imprisonment. The cap on administrative or civil penalties should be substantially reduced and should be commensurate to the nature of the violation, although we support the minimum penalty of $250. The decreased penalties clause, Subsection 7(c), should be made mandatory.

Finally, we support Subsection 7(e), pertaining to revocation of excavator licenses, because habitual violators endanger themselves and others and give our industry a public relations black eye. The term "habitual violator" is vague and should be defined. For example, four violations in three years might constitute a habitual violator. In addition, a habitual violator may be a governmental entity. We recommend that governmental entities that are habitual violators be forced to privatize their excavation activities.

Section 8. Department of Transportation

NUCA supports this Section, especially the proposed model state program and public education campaign. Again, we are pleased that the Secretary of Transportation will develop these programs in consultation with excavators, facility operators, and one-call system operators.

Section 9. Alternate Form of State Program

We have no problem supporting this section so long as all facility operators are covered by alternative state programs.

Section 10. State Reports

We support this section. The state reports should include information pertaining to the nature and frequency of underground facility accidents caused by third-party damage and other factors.

Public Service Commission of the District of Columbia

450 5TH STREET, N.W.
WASHINGTON, D.C. 20001
(202) 626-5100

IN REPLY REFER TO:

July 28, 1994

Honorable Phillip R. Sharp
Chairman
Subcommittee on Energy and Power
Committee on Energy and Commerce
U.S. House of Representatives
Washington, DC 20515-6117

Dear Chairman Sharp:

The following comments are in response to your letter of July 15, 1994 which posed five follow-up questions to the June 23 hearing at which I appeared as a witness for the National Association of Pipeline Safety Representatives (NAPSR). I appreciate the opportunity to respond on behalf of NAPSR and compliment you and the Committee on your forthright and efficient handling of the one-call issue.

Each question is restated below along with my comments thereon:

1) **Should there by any exceptions to universal, mandatory participation in a one-call system? Do you think that exempting gathering lines and flow lines undermines our goal of universal, mandatory participation?**

Section 5 (b) of H.R. 4394 provides adequate flexibility for the few exemptions from one-call participation that may be necessary. It provides that for "... excavation on property owned by that person, or for the purposes of excavation under the exclusive control of that person, and the same person operates all underground facilities at the site of the excavation, a State program need not require that such person contact the one-call notification system before excavating."

This allows for the individual states to make determinations on reasonable exemptions to one-call requirements. It obviously makes little sense to require cemeteries, tree nurseries, or mining companies to call one-call for each excavation in dedicated work areas. Similarly, flow lines

Hon. Phillip R. Sharp
July 28, 1994
Page 2 of 3

which are located on gas or oil lease properties appear to fall under the exemptions contemplated under Section 5 (b), although the language indicates only "owners" of property can enjoy exemptions. Perhaps this should be expanded to include "owners and operators."

However, state departments of transportation, local departments of public works, and municipal utilities, all of which routinely perform excavation in public rights-of-way, should never be exempted from mandatory one-call requirements. To exempt these entities would reduce the effectiveness of one-call systems to prevent underground excavation accidents. The same argument applies to oil and gas gathering lines which traverse numerous public and private properties.

2) **What are the essential elements that should be included in Federal legislation as minimum requirements of a one-call system? (Please refer to my discussion draft of the amendment in the nature of a substitute to H.R. 4394.)**

In general, the elements included in the proposed legislation as minimum requirements for one-call systems are adequate.

However, the increasing use of directional boring for installation of distribution piping requires that some protection needs to be assured for using this procedure. The federal legislation might require underground facility owners to verify the depth of underground facilities prior to directional boring.

3) **How important is the role of pubic education in the use of one-call systems? Do you have any suggestions, other than the use of public service announcements which has been provided for in the bill, on how we can best communicate the importance of this issue to the public?**

Pubic education is very important in assuring compliance with one-call provisions. However, the burden of developing public information need not and should not be forced on federal agencies. Existing one-call systems are actively engaged in public information activities, and do not need federal materials or involvement. In addition, we do not know of any public information techniques that are not currently being used by one-call operations throughout the country.

4) **Do yo think the penalty provisions in H.R. 4394 are too high? What type of penalty structure should be included in Federal Legislation?**

The penalty provisions in Section 6 of H.R. 4394 are appropriate.

5) **Do you agree or disagree with Chair Moler's assessment of siting provisions referred to in H.R. 4616? Do you think that FERC should have a greater role in considering safety when siting pipelines?**

We believe FERC has adequate authority to consider safety when siting interstate pipelines, but do not believe that FERC should promulgate safety rules or enforce safety provisions through its siting process. A good way for FERC to take safety into consideration, without "getting into the safety business," is for it to request comments from the Office of Pipeline Safety, state agencies, and other safety responsibility centers concerning interstate pipeline siting proposals.

For pipelines for which siting is not regulated by the FERC, route regulation is a function best left to the states, as they are more cognizant of local conditions. Indeed, states have been carrying this burden in siting cases for many years.

Finally, there is no conflict between the specific safety standards of 49 CFR 192 and 195 and safety as a routing issue. OPS has safety standards addressing the manner of construction of lines in developed areas, but whether the line should be placed in a particular developed area at all is a public policy issue. Federal law should not prohibit states from regulating routing.

If you have any questions or need clarification of any of the above items, please contact me at (202) 626-5156.

Sincerely,

Richard D. Huriaux, P.E.
National Association of
Pipeline Safety Representatives

cc: Mary McDaniel
President, NAPSR

J. Bell
A. Jaiyeola